Álvaro Herrero and Emilio Corchado

Mobile Hybrid Intrusion Detection

Studies in Computational Intelligence, Volume 334

Editor-in-Chief
Prof. Janusz Kacprzyk
Systems Research Institute
Polish Academy of Sciences
ul. Newelska 6
01-447 Warsaw
Poland
E-mail: kacprzyk@ibspan.waw.pl

Further volumes of this series can be found on our homepage: springer.com

Vol. 312. Patricia Melin, Janusz Kacprzyk, and Witold Pedrycz (Eds.)
Soft Computing for Recognition based on Biometrics, 2010
ISBN 978-3-642-15110-1

Vol. 313. Imre J. Rudas, János Fodor, and Janusz Kacprzyk (Eds.)
Computational Intelligence in Engineering, 2010
ISBN 978-3-642-15219-1

Vol. 314. Lorenzo Magnani, Walter Carnielli, and Claudio Pizzi (Eds.)
Model-Based Reasoning in Science and Technology, 2010
ISBN 978-3-642-15222-1

Vol. 315. Mohammad Essaaidi, Michele Malgeri, and Costin Badica (Eds.)
Intelligent Distributed Computing IV, 2010
ISBN 978-3-642-15210-8

Vol. 316. Philipp Wolfrum
Information Routing, Correspondence Finding, and Object Recognition in the Brain, 2010
ISBN 978-3-642-15253-5

Vol. 317. Roger Lee (Ed.)
Computer and Information Science 2010
ISBN 978-3-642-15404-1

Vol. 318. Oscar Castillo, Janusz Kacprzyk, and Witold Pedrycz (Eds.)
Soft Computing for Intelligent Control and Mobile Robotics, 2010
ISBN 978-3-642-15533-8

Vol. 319. Takayuki Ito, Minjie Zhang, Valentin Robu, Shaheen Fatima, Tokuro Matsuo, and Hirofumi Yamaki (Eds.)
Innovations in Agent-Based Complex Automated Negotiations, 2010
ISBN 978-3-642-15611-3

Vol. 320. xxx

Vol. 321. Dimitri Plemenos and Georgios Miaoulis (Eds.)
Intelligent Computer Graphics 2010
ISBN 978-3-642-15689-2

Vol. 322. Bruno Baruque and Emilio Corchado (Eds.)
Fusion Methods for Unsupervised Learning Ensembles, 2010
ISBN 978-3-642-16204-6

Vol. 323. Yingxu Wang, Du Zhang, and Witold Kinsner (Eds.)
Advances in Cognitive Informatics, 2010
ISBN 978-3-642-16082-0

Vol. 324. Alessandro Soro, Vargiu Eloisa, Giuliano Armano, and Gavino Paddeu (Eds.)
Information Retrieval and Mining in Distributed Environments, 2010
ISBN 978-3-642-16088-2

Vol. 325. Quan Bai and Naoki Fukuta (Eds.)
Advances in Practical Multi-Agent Systems, 2010
ISBN 978-3-642-16097-4

Vol. 326. Sheryl Brahnam and Lakhmi C. Jain (Eds.)
Advanced Computational Intelligence Paradigms in Healthcare 5, 2010
ISBN 978-3-642-16094-3

Vol. 327. Slawomir Wiak and Ewa Napieralska-Juszczak (Eds.)
Computational Methods for the Innovative Design of Electrical Devices, 2010
ISBN 978-3-642-16224-4

Vol. 328. Raoul Huys and Viktor K. Jirsa (Eds.)
Nonlinear Dynamics in Human Behavior, 2010
ISBN 978-3-642-16261-9

Vol. 329. Santi Caballé, Fatos Xhafa, and Ajith Abraham (Eds.)
Intelligent Networking, Collaborative Systems and Applications, 2010
ISBN 978-3-642-16792-8

Vol. 330. Steffen Rendle
Context-Aware Ranking with Factorization Models, 2010
ISBN 978-3-642-16897-0

Vol. 331. Athena Vakali and Lakhmi C. Jain (Eds.)
New Directions in Web Data Management 1, 2011
ISBN 978-3-642-17550-3

Vol. 332. Jianguo Zhang, Ling Shao, Lei Zhang, and Graeme A. Jones (Eds.)
Intelligent Video Event Analysis and Understanding, 2011
ISBN 978-3-642-17553-4

Vol. 333. Fedja Hadzic, Henry Tan, and Tharam S. Dillon
Mining of Data with Complex Structures, 2011
ISBN 978-3-642-17556-5

Vol. 334. Álvaro Herrero and Emilio Corchado
Mobile Hybrid Intrusion Detection, 2011
ISBN 978-3-642-18298-3

Álvaro Herrero and Emilio Corchado

Mobile Hybrid Intrusion Detection

The MOVICAB-IDS System

Springer

Dr. Álvaro Herrero
University of Burgos
Civil Engineering Department Polytechnic School
Francisco de Vittoria s/n
09006 Burgos
Spain
E-mail: ahcosio@ubu.es

Prof. Dr. Emilio Corchado
University of Salamanca
Departamento de Informáca y Automtica
Facultad de Biología
Plaza de la Merced s/n
37008 Salamanca
Spain
E-mail: escorchado@usal.es

ISBN 978-3-642-18298-3 e-ISBN 978-3-642-18299-0

DOI 10.1007/978-3-642-18299-0

Studies in Computational Intelligence ISSN 1860-949X

Typeset & Cover Design: Scientific Publishing Services Pvt. Ltd., Chennai, India.

Printed on acid-free paper

9 8 7 6 5 4 3 2 1

springer.com

Contents

Abbreviations

AI	Artificial Intelligence.
ANN	Artificial Neural Network.
BDI	Belief, Desire and Intention.
CBR	Case-Based Reasoning.
CCA	Curvilinear Component Analysis.
CMLHL	Cooperative Maximum-Likelihood Hebbian Learning.
EPP	Exploratory Projection Pursuit.
HIDS	Host-Based Intrusion Detection System.
ID	Intrusion Detection.
IDS	Intrusion Detection System.
MAS	Multiagent System.
MIB	Management Information Base.
MLHL	Maximum-Likelihood Hebbian Learning.
NFN	Negative Feedback Network.
NID	Network-Based Intrusion Detection.
NIDS	Network-Based Intrusion Detection System.
PCA	Principal Component Analysis.
SNMP	Simple Network Management Protocol.
SOM	Self-Organizing Map

Preface

This monograph gathers research efforts performed over a period of about five years and comprises works on network-based Intrusion Detection (ID) that is grounded on visualisation and hybrid Artificial Intelligence (AI). It has led to the design of MOVICAB-IDS (MObile VIsualisation Connectionist Agent-Based IDS), a novel Intrusion Detection System (IDS), which is comprehensively described in this book.

This novel IDS combines different AI paradigms to visualise network traffic for ID at packet level. It is based on a dynamic Multiagent System (MAS), which integrates an unsupervised neural projection model and the Case-Based Reasoning (CBR) paradigm through the use of deliberative agents that are capable of learning and evolving with the environment. The proposed IDS applies a neural projection model to extract interesting projections of a traffic dataset and to display them through a mobile visualisation interface. As a result of depicting each simple packet and preserving the temporal context, MOVICAB-IDS provides security personnel with a synthetic, intuitive snapshot of network traffic and protocol interactions. This visualisation interface supports the straightforward detection of anomalous situations and their subsequent identification. Additionally, it helps ascertain the internal structure and the behaviour of the traffic data, thereby improving supervision of network activity.

The performance of MOVICAB-IDS was tested in different domains which entailed several attacks and anomalous situations and was further verified through a two-fold analysis. The proposed IDS was validated with a novel mutation-based testing method especially developed for that purpose, and the projections of its underlying neural model were compared with those obtained with some other projection models.

The monograph subsumes research results of the authors, a large part of which comes from Álvaro Herrero's PhD dissertation prepared at the University of Burgos (Spain) under the supervision of Dr. Emilio Corchado.

May 2010

Álvaro Herrero and Emilio Corchado
Burgos and Salamanca
Spain

Chapter 1
Introduction

The present book proposes, describes, and tests a novel Intrusion Detection System (IDS) called **MOVICAB-IDS** (MObile VIsualisation Connectionist Agent-Based IDS) that focuses on network-based Intrusion Detection (ID) from the visualisation and hybrid Artificial Intelligence (AI) standpoints. The proposed IDS combines different AI paradigms to visualise network traffic for ID at packet level. It is based on a MAS, which integrates an unsupervised neural projection model and the Case-Based Reasoning (CBR) paradigm through the use of deliberative agents that are capable of learning and evolving with the environment. By means of the applied neural projection model, MOVICAB-IDS extracts interesting projections of a traffic dataset and displays them through a mobile visualisation interface. As a result of depicting each simple packet and preserving the temporal context, MOVICAB-IDS is able to provide security personnel with a synthetic, intuitive snapshot of network traffic and protocol interactions. This visualisation interface supports the straightforward detection of anomalous situations and their identification. Additionally, it can help to ascertain the internal structure and behaviour of the traffic data, thereby improving supervision of network activity.

A dataset of network traffic data (GICAP-IDS) has been generated for the experimental study of the proposed IDS. As MOVICAB-IDS has been designed for the effective detection of SNMP and probing attacks, instances of such attacks are contained in the generated dataset. These attacks were created over a background of "normal" traffic in a previously studied network. The performance of MOVICAB-IDS has been checked in different domains containing several attacks and anomalous situations.

To assess MOVICAB-IDS, a novel testing method based on mutations was developed and applied to the proposed IDS. The main idea behind this technique is to confront MOVICAB-IDS (and some other IDSs) with previously unseen attacks. These novel situations simulate the new attacks (know as "0-day" attacks) that a computer system may face. Additionally, the neural model supporting the visualisation capabilities of MOVICAB-IDS is tested by comparing its projections to those generated by some other neural projection models.

Á. Herrero and E. Corchado: Mobile Hybrid Intrusion Detection, SCI 334, pp. 1–2.
springerlink.com

The organisation of this book is outlined in the following paragraphs.

Chapter 2 introduces the field in which the present study is located, namely Intrusion Detection (ID) within the framework of Computer Security. The main paradigms supporting the proposed IDS (Artificial Neural Networks, Multiagent Systems, and Case-Based Reasoning) are also introduced in this chapter.

Chapter 3 contains a state-of-the-art review of research work on Network-Based ID (NID). It presents the most relevant advances in applying the perspectives discussed in this book. Previous work on visualisation for NID is described in the context of current visualisation techniques and the visualised data. Moreover, agent-based IDSs, in general, and mobile agent-based IDSs, in particular, are comprehensively described in this chapter. Finally, the main novelties of MOVICAB-IDS in the light of these previous works are reported and discussed.

Chapter 4 provides a general overview and a detailed description of the IDS proposed in this book: MOVICAB-IDS. Information on the methodology followed in the development of such system is also included in this chapter. Each agent in the MAS design and their inner structure are described in this chapter.

The experimental results of applying MOVICAB-IDS to the GICAP-IDS and to the 1998 DARPA dataset [1] are presented and explained in Chapter 5. Both datasets are described, although preferential attention is given to the GICAP-IDS dataset which was specifically generated for this study. Finally, the main features of the visualisations obtained with the two datasets are highlighted.

Chapter 6 presents the performance analysis of the proposed IDS. A novel testing technique based on mutations is introduced and subsequently applied to MOVICAB-IDS. A two-fold analysis is performed by applying the testing technique to a sample dataset and accumulated segments. Additionally, a comparative study involving different well-known projection models is reported.

The results detailed in the two aforementioned chapters are discussed in Chapter 7, which also puts forward a number of global conclusions drawn from the study as a whole. This chapter also sets out pointers for future work on ways to improve the proposed IDS.

Chapter 2
Visualisation, Artificial Intelligence, and Security

As is well known, the rapid growth of computer networks and their interconnections have and will continue to entail security problems. New security failures are discovered on a daily basis and a growing number of hackers are trying to take advantage of such failures. Consequently, many organizations invest huge amounts of time, effort, and resources in security for information systems, due to the need to protect their systems from these intruders. Some background information on the general field of computer systems security is provided in section 2.1.

Intrusion Detection (ID) is a subfield of computer system security that focuses on the identification of attempted or ongoing attacks on a computer system or network. Detailed information about IDSs is included in section 2.2. As this work proposes visualisation-based ID, a simple introduction to visualisation for network security is given in 2.3, before introducing the most common visualisation techniques in section 2.4. This work describes a novel IDS that approaches ID from the standpoint of hybrid AI, by combining the following AI paradigms:

- **Artificial Neural Network** (ANN), introduced in section 2.5.
- **Multiagent System** (MAS), introduced in section 2.6.
- **Case-Based Reasoning** (CBR), introduced in section 2.7.

2.1 Computer System Security

Faced with rising rates of computer crime, the protection of computer systems has become an extremely critical activity. It is enough to examine the following statistics to understand the extent of this problem:

- Data from CERT [2] show steep increases in computer crime concerning communications, which rose from 6 reported incidents in 1988 to 137,529 incidents in 2003. More up-to-date statistics on these incidents are not available from the CERT website as "*attacks against Internet-connected systems have become so commonplace that counts of the number of incidents reported provide little information with regard to assessing the scope and impact of attacks*" [2].

Á. Herrero and E. Corchado: Mobile Hybrid Intrusion Detection, SCI 334, pp. 3–39.
springerlink.com

- As reported in [3], Symantec created 1,656,227 new malicious code signatures in 2008. This is a 165% increase over those created in 2007, when 624,267 new malicious code signatures were added. In other words, "*a significant spike in new malicious code threats occurred during 2008*" [3].
- The 2008 CSI Computer Crime and Security Survey [4] ("*perhaps the most widely quoted set of statistics in the industry*") is based on the responses of 522 computer security practitioners in U.S. corporations, government agencies, financial institutions, medical institutions and universities. It reported an average (per respondent) economic loss caused by various types of computer security incidents of $288,618, down from $345,005 in the 2007 survey [5], but up from the low of $167,713 reported in 2006. For the 2008 survey, only 144 respondents answered questions on economic loss.
- Some of the major security breaches that became public in the fourth quarter of 2008, as reported by IBM X-Force [6], were:
- Account information of 160,000 users of CheckFree was compromised due to a domain hijacking.
- T-Mobile acknowledged a breach that occurred in 2006 exposing records of 17 million German customers.
- A laptop containing private information on 97,000 Starbucks employees was stolen.
- A data breach at the University of Florida exposed personal records of 330,000 patients of its College of Dentistry.
- Private records of 1.5 million clients of RBS WorldPay were exposed due to a system compromise.
- There is a lack of security statistics in the European context, but according to [7], "*spending on IT security services in Western Europe will reach $7.5 billion by 2009*".

As a society, we have become extremely dependent on the use of information systems, so much so that the danger of serious disruption of crucial operations is frightening [8]. Thus, there is an unquestionable need for organisations to protect their systems from intrusions and consequently, new network security tools are being developed.

The ever-changing nature of attach technologies increases the vulnerability of computers and networks, which further complicates the protection of such systems. This can be seen by comparing the evolution of computer misuse over the past twenty years [9]. The progressive evolution of attacks could mainly be attributed to the dynamic nature of systems and networks, the creativity of attackers and the wide range of computer hardware and software. Such complexity arises when dealing with distributed network-based systems and insecure networks such as the Internet. A graphical representation of the evolution of attacks from [10] is shown in Fig. 2.1.

As can be seen in Fig. 2.1, there is an inverse relationship between the decreasing technical knowledge required to execute attacks and the increasing sophistication of those attacks. In other words, "*less skill is needed to do more damage*" [11].

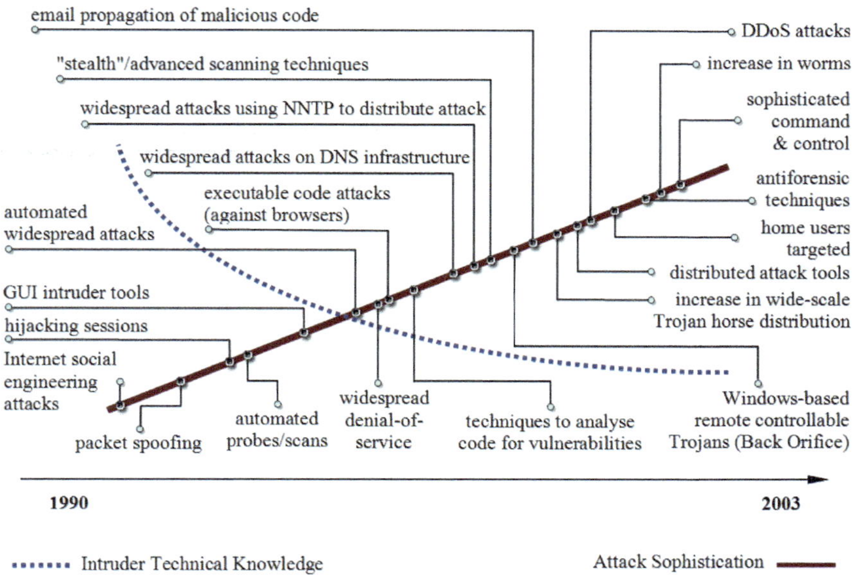

Fig. 2.1 Evolution of Attack Sophistication/Intruder Knowledge (source: CERT).

Any attack on or intrusion into a system is likely to affect at least one of the three principles of information security: integrity, confidentiality, and availability of information. Thus, an intrusion can be generally defined as "*any set of actions that attempt to compromise the integrity, confidentiality or availability of a resource*" [12], where:

- **Integrity** refers to the goal of keeping trustworthy information free from undesired change, whether malicious or benign [13].
- **Availability** refers to the goal of keeping resources available for authorised users [13].
- **Confidentiality** refers to the goal of keeping sensitive information private and accessible only to certain people [13].

In the past, computer and network security relied on creating completely secure systems to preserve these principles of information security. This is no longer a valid approach, mainly because of the following reasons [14]:

- Software often contains flaws that may create security problems [13, 14] and furthermore, software upgrades may also introduce new problems. According to [15], there are two main reasons for that: "*most designers and developers are not trained in general security principles and market forces dictate that ever more complex software products are delivered at accelerated speeds*".
- The increasing demand for network connectivity makes it difficult, if not impossible, to isolate a system so as to protect it from external penetration.

- A central component of computer systems, the computer network itself, may not be secure enough. Experts have acknowledged a number of inherent security flaws in the widely-used TCP/IP suite (regardless of its particular implementation).

The above-mentioned flaws are known as system vulnerabilities, regardless of the component they refer to (software, operating systems, protocols, etc.). The vulnerabilities that can be exploited by attackers can then range from a flaw in a piece of software (e.g. a buffer overflow that can be exploited to elevate privileges) to a flaw in an organisational structure that allows a social engineering attack to obtain sensitive information or account passwords [16]. A detailed list of the more common vulnerability types can be found in [17]. It is enough to look at the following figures, to gain an idea of the vulnerability problem and its magnitude. Symantec documented 5,491 new vulnerabilities in 2008 [3], 2% of which it rated as high severity, while 68% were rated as medium severity, and 30% as low severity. The problem is not only the existence of these vulnerabilities but also their removal. According to [18], only 47% of all vulnerabilities disclosed in 2008 would be corrected through vendor patches, while the remaining ones might never even be patched.

After realising that a perfectly secure system can not be built, we should therefore expect systems to fail and be prepared for such failures [19]. Thus, we should be aware that protection and attack-prevention mechanisms are not enough in themselves. As stated in [20]: "*detection and monitoring will always be needed, no matter what the advances in prevention the future might bring*". In response to the above mentioned difficulties in developing a fully secure system, a more realistic approach to security management has been established in recent years. This approach divides the field of Computer System Security into four non-trivial, challenging, but more manageable sub-problems (see Fig. 2.2).

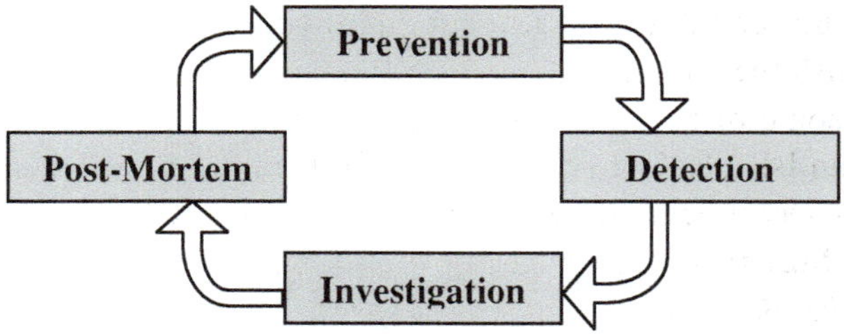

Fig. 2.2 A Computer System Security Management Model (from [14]).

Under this approach, developing secure systems only represents one of the four components of the security management model; the prevention component in Fig. 2.2. The other three components can be briefly defined as:

- **Detection component:** identifies security breaches that are or can be exploited.
- **Investigation component:** determines exactly what happened on the basis of data from the detection component. This component may also include the gathering of further data in order to identify the security violator.

- **Post-mortem component:** analyses how to prevent similar intrusions in the future. It comprises the extraction, documentation, examination, preservation, analysis, evaluation, and interpretation of materials to provide relevant and valid information as evidence in civil, criminal or administrative proceedings [21]. In general, computer forensics looks for digital evidence and examines what can be retrieved from storage media [21].

Some other authors simplify this system security management model by splitting it into three components: **protection**, **detection**, and **response**. The investigation and post-mortem components are included in the response component of this simpler model. Responses may vary between mitigating the current situation, analysing the incident, repairing any damage, and improving the protection/detection mechanisms to prevent similar occurrences in the future.

In the past, the attention of computer security researchers centred mainly on the prevention component, which encompasses a wide range of techniques and tools: antivirus software, firewalls, cryptography, etc. Nevertheless, these tools can not fully authenticate or completely block certain services, as these are usually made available to everyone connected to the Internet. Given that prevention is not enough for network security, it is necessary to install an additional defence to receive early warnings on malicious activity and to counter intrusions when the first perimeter of defence has been penetrated. This is the main goal of the detection component.

The detection component must assume that an attacker can obtain access to the desired targets and will at some point successfully violate a given security policy. When that occurs, this component must report that something is wrong. The most advanced systems will also react in an appropriate way (investigation and post-mortem/reaction component).

The emergence and the proven utility of ID have meant that greater attention is now paid to the detection component. IDS are the most important tools in the detection component and have become increasingly popular over recent years, although firewalls are still the most widely used tool in the security field [22]. The 2008 CSI Computer Crime and Security Survey [4] reported that 69% of the surveyed organisations included IDSs in their security infrastructure. Complementing IDSs, there are some other tools for the detection component such as System Integrity/Vulnerability Verifiers, Log File Monitors, File Integrity Checkers, Honeypots and Padded Cells. Further information on these other tools may be found in [17, 21].

As part of the detection component, network monitoring attempts to detect attacks on a network or on the network hosts, by monitoring the network traffic [23]. In other words, network monitoring tools supervise the traffic in computer networks and generate alerts, or trigger defensive actions whenever suspect activities are detected.

The research work described in this book centres on enhancing the detection component. It is beyond the scope of this section to provide further details on the great amount of network security vulnerabilities, strategies, mechanisms, tools and so on. Thus, following a general overview of network security, the next section will examine the topic of IDSs.

2.2 Intrusion Detection Systems

Plenty of published studies survey different issues in the field of ID [31, 36, 37, 40, 92, 164, 169, 176, 196, 205, 216, 236] such as its history, approaches, techniques, tools, taxonomies, and so on. As comprehensive information on ID can be found in the referenced works and in many others, this section provides a brief overview of the field, introducing the main concepts and background knowledge.

As previously stated, an intrusion can be defined as a set of actions that attempt to compromise any of the three security principles (integrity, confidentiality, and availability) of any resource in a computing system by exploiting vulnerabilities [43, 171, 243]. In other words, these are actions that attempt to compromise a target system, regardless of whether they are successful.

Intrusions can be produced by attackers that access the system, by authorized users that attempt to obtain unauthorized privileges, or by authorized users that misuse the privileges given to them. The complexity of such situations increases in the case of distributed network-based systems and insecure networks. When attackers try to access a system through external networks such as the Internet, one or several hosts may be involved. In some cases, the attacker may make use of remote machines that were targeted in earlier attacks to carry out the attack. Nevertheless, an intrusion can also involve numerous intruders that target various victim systems. From a victim's perspective, intrusions are characterized by their manifestations, which might or might not include damage [24]. Some attacks may produce no manifestations while some apparent manifestations can be produced by system or network malfunctions.

The following scenarios are examples of intrusions [14]:

- An employee browses through employee reviews without permission.
- A user exploits a flaw in a file server program to gain access to it and then to corrupt another user's file.
- A user exploits a flaw in a system program to obtain super-user status.
- An intruder uses a script to "crack" the passwords of other users on a computer.
- An intruder installs a "snooping program" on a computer to inspect network traffic, which often contains user passwords and other sensitive data.
- An intruder modifies the router tables in a network to prevent the delivery of messages to a particular computer.
- A worm automates attacks originating from other computers.

According to [25], intrusions can be divided into the following six main types, although some other authors [26] propose a different taxonomy.

- **Attempted break-ins:** an outsider "convinces" the system that an authorized user has logged on by supplying a valid user identification and password.
- **Masquerade attacks:** the intruder attempts to "convince" the system that the user it expects to log on is a different user, presumably with higher privileges.
- **Penetration of the security control system:** a user attempts to modify the security characteristics of the system such as its passwords and authorizations.
- **Leakage:** causes information to move out of the system.

- **Denial of service:** making system resources unavailable to other users.
- **Malicious use:** resource hogging, file deletion, and other miscellaneous attacks.

Despite the fact that there is a wide range of intrusion goals and types, a general five step intrusive process is defined [27] as:

- **Reconnaissance:** before launching an attack, attackers conduct detailed reconnaissance to collect information on their victims.
- **Scanning:** the attacker, equipped with information on the infrastructure of the victim's system, begins scanning it to look for vulnerabilities and openings.
- **Gaining access:** if the attacker is a legitimate user of the system, then the attack will probably attempt to gain access through the operating system and the application. If the attacker is an outsider, then the attack is most likely to be launched through the network.
- **Maintaining access:** having gained access to the target system, the attacker needs to maintain it.
- **Covering tracks:** just before the attack is over, experienced hackers will try to cover their tracks so that the attack goes unnoticed. Beyond that, attackers will also attempt to modify system logs and create covert channels to transmit data without the victim's knowledge.

We may examine two different types of intrusions by considering the source of the intrusion: internal (the intruder is an authorized user of the system) and external (the intruder is not a user of the system). The five-step intrusion process defined above may vary depending on the intrusion source.

An IDS can be defined as a piece of software that runs on a host, which monitors the activities of users and programs on the same host and/or the traffic on networks to which that host is connected [27]. The main purpose of an IDS is to alert the system administrator to any suspicious and possibly intrusive event taking place in the system that is being analysed. Thus, they are designed to monitor and to analyse computer and/or network events in order to detect suspect patterns that may relate to a system or network intrusion.

Ever since the first studies in this field [28, 29] in the 80s, the accurate detection of computer and network intrusions in real-time has been an interesting and intriguing problem for system administrators and information security researchers. The first paper on ID [28] stated that a certain class of intrusions could be automatically identified by analysing computer audit trails. Almost 30 years since this paper was published, there is still no such thing as a 100% effective IDS. People working in this area have concluded that traffic analysis and ID in large networks is a highly complex task, which is mainly due to the following reasons: the intensity and the heterogeneous nature of the traffic phenomenon, the dynamic nature of systems and networks, the creativity of attackers, and the wide range of computer hardware and operating systems, among others. According to [30], IDS designers usually make a set of assumptions to limit the scope of the problem of detecting intrusions:

- Total physical destruction of the system (the ultimate denial of service) is not considered. IDSs are usually based on the premise that the operating system

and the IDS will continue to function for at least some period of time. Hence, it can alert administrators and support subsequent actions.
- ID is not a problem that can be solved from time to time. As a consequence, continual vigilance is required.
- Vulnerabilities are usually assumed to be independent. Even after a known vulnerability has been removed, a system administrator may run an IDS in order to detect attempts at penetration, even though they are guaranteed to fail. Most IDSs do not take into account whether specific vulnerabilities have been fixed or not.

IDSs are increasingly being viewed as an important component in a comprehensive security solution, and companies that build security components are integrating them into comprehensive system management tools [31]. Thus, IDSs are common elements in modern infrastructures that enforce network policies. Today's commercial systems typically rely on a knowledge base of rules to discriminate normal from malicious traffic. The set of rules, however, is susceptible to inconsistencies, and continuous updating is required to cover previously unseen attack patterns.

According to [32], there are some features that can be analysed when considering IDSs:

- **Accuracy:** an IDS must not identify a legitimate action in a system environment as an intrusion, an event referred to as a false positive.
- **Completeness:** an IDS should not fail to detect an intrusion. An unnoticed intrusion is referred to as a false negative.
- **Performance:** is the rate at which data are processed by an IDS. The performance of an IDS must be sufficiently good to carry out real time ID. Here, real time means that an intrusion has to be detected before significant damage has been caused.
- **Fault Tolerance:** an IDS must itself be resistant to attacks. As explained in [16], aware that an IDS has been deployed on a network, some intruders are likely to attack the IDS first, to disable it or to force it to provide false information. Very few papers [33, 34] have made an effort to identify and to fix IDS-related vulnerabilities.
- **Timeliness:** an IDS must perform and propagate its analysis as quickly as possible so as to enable a reaction before too much damage is inflicted.

The accuracy and completeness features are related to false alarms which are classified as being either false positive or false negative. A false positive involves a legitimate event being reported by the IDS as an intrusion. On the contrary, a false negative involves an intrusion that goes undetected. As in some other fields, false negatives are the most dangerous although false positives are also harmful.

2.2.1 A General Architecture for ID

A general IDS architecture [35] that has been widely used consists of five main components: data gathering device (or sensor), detector (ID analysis engine), knowledge base (database), configuration device, and response component. However, certain IDSs do not fit into that structure, which is the case of the IDS proposed in this book. It applies a more general architecture adapted from [16] that logically organises the functionality of an IDS into three components:

- **Sensors:** in charge of collecting data. The input for a sensor may be any part of a system that could contain evidence of an intrusion. Sensor inputs can range from network packets or flow data to system call traces or log files.
- **Analysers:** in charge of determining if an intrusion has occurred. The input of this component is the data captured by the sensors, and the output is an indication that an intrusion has occurred.
- **User interface:** enables users to view the output of the IDS and, in some cases, to control the behaviour of the system.

2.2.2 IDS Taxonomy

IDSs may be classified on the basis of various different features [31, 39, 208], although only some of the most characteristic ones [32] will be used in this section. Subsets of these features, which refer to the internal workings of the IDS can be defined: detection method, monitoring scope, and behaviour on detection.

A standard characterization of IDSs, based on their **detection method**, or model of intrusions, defines the following paradigms:

- **Anomaly-based ID** (also known as behaviour-based ID)**:** the IDS detects intrusions by looking for activity that differs from the normal behaviour that is associated with the user or the system. To do so, the observed activity is compared against previously "defined" profiles of expected normal usage. These profiles may refer to users, groups of users, applications, or system resource usage [21].
  In anomaly detection, it is assumed that all intrusive activities are necessarily anomalous. Unfortunately, in some cases, instead of their being identical, the set of intrusive activities only intersects the set of anomalous activities. As a consequence, two main problems arise [36]:
  - Anomalous activities that are not intrusive are flagged as intrusive (i.e. false positives).
  - Intrusive activities that are not anomalous are not flagged up (i.e. false negatives).

Anomaly-based IDSs can support time-zero detection of novel attack strategies but may suffer from a relatively high rate of false positives [37]. The main issues in these IDSs are the selection of threshold levels to reduce the two aforementioned problems and the selection of features to monitor the traffic [36].

- **Misuse-based ID** (also known as knowledge-based ID)**:** detects intrusions by looking for activity that corresponds to known intrusion techniques (signatures) or system vulnerabilities. Misuse-based IDSs are therefore commonly known as signature-based IDSs. They detect intrusions by exploiting the accumulated knowledge on specific attacks and vulnerabilities.

As opposed to anomaly detection, misuse detection assumes that each intrusive activity can be represented by a unique pattern or signature [21]. Slight variations of the same intrusive activity will then produce new signatures. This approach entails one main problem; intrusions whose signatures are not archived by the system can not be detected. As a consequence, a misuse-based IDS will never detect a new (previously unseen) attack [21], also known as 0-day attack. The completeness of such IDSs requires regular updating of their knowledge of attacks.

The main issues in misuse-based IDSs are how to write a signature that encompasses all possible variations of the attack in question, and how to write signatures that do not also match non-intrusive activity [36].

- **Specification-based ID:** this third paradigm was introduced in [38]. It relies on program behavioural specifications reflecting system policies that are used as a basis to detect attacks. This technique has been proposed as a promising alternative that combines the strengths of the two previous ones.

Characterization of IDSs based on their **monitoring scope** (also known as audit source location or data source):

- **Host-based IDS (HIDS):** audit data from a single host is used to detect intrusions. A HIDS is deployed on a single target computer and usually monitors operating system logs to detect sudden changes in these logs. They may include system, event, and security logs on Windows systems and syslog in Unix environments [21].
- **Multihost-based IDS:** audit data from multiple hosts is used to detect intrusions. Hosts in a network segment can be monitored by these IDSs.
- **Network-based IDS (NIDS):** network traffic data is used to detect intrusions. The IDS captures and inspects (depending on the technology) every packet sent to the network under analysis, regardless of whether it is permitted by a firewall [21]. Interesting issues about where to place NIDS sensors are discussed in [21, 39].

  Among the main problems for NIDSs are the ability of a skilled attacker to evade detection by exploiting ambiguities in the traffic stream (as perceived by the NIDS [33]) and the handicap of analysing encrypted data. On the contrary, such IDSs hold several advantages over HIDSs [35, 40]:

  - They can analyse network-level events.
  - They can be installed in a network without affecting the existing systems that are already monitored.
  - They monitor an entire network segment simultaneously.
  - NIDSs are usually more resistant, as they do not have to reside on the host that is targeted by the attack to be identified.
  - They do not depend on the operating systems installed in the hosts.

- **Host/network-based IDS:** network traffic data, along with audit data from several hosts, is used to detect intrusions. Designed by combining HIDS and NIDS approaches, this category is meant to make the most of the advantages of these two approaches. One of the key issues of this fusion approach is the development of an interface that can show all the distributed information captured by the set of sensors.

This characterization of IDSs based on their scope reflects the history of computing [32]. When the first IDSs were designed, the target environment was a mainframe computer and all users were local to the system under analysis. Thus, initial IDSs were conceived as host-based. NIDSs emerged as the focus of computing shifted from mainframe environments to distributed networks of hosts.

Characterization of IDSs based on their **behaviour**:

- **Passive:** when an intrusion is detected, an alert is generated without trying to oppose the attack any further: passive IDSs simply report an event, but no countermeasure is applied to stop the attack.
- **Active:** they can change the security posture of a protected system to react to the detected intrusion, which is an automatic response. As an example, once an intrusion has been detected, these IDSs may modify file permissions, add firewall rules, kill processes, inject TCP reset packets to cut suspicious connections, and reconfigure routers and firewalls to block packets from the attacker's apparent location (IP address) [17, 41]. For instance, one issue that should be taken into account in the case of a false positive is that automated responses can lead to a denial of service to legitimate users of the system [38].

Characterization of IDSs based on their usage frequency [41]:

- **On-line analysis:** real-time analysis of the activity of the system to be protected. Data is examined as soon as it is produced and alarms must be raised as soon as an attack is detected, so that immediate action may be taken against it.
- **Off-line analysis:** analyses a snapshot of the system state and assesses its security. By running only occasionally, they may perform a more thorough analysis without having an unacceptable impact on the performance of the monitored system.

There are many examples of IDSs for each one of the categories defined in this section. A description and review of some of these examples is provided in Chapter 3.

2.3 Visualisation for Network Security

Visualisation techniques have been applied to massive datasets for many years. They are considered a viable approach to information seeking, as humans are able to recognize different features and to detect anomalies by inspecting graphs [42]. Visual inspection of network traffic patterns is presented as an alternative for managing a crucial aspect of network monitoring [43], as its chief aim is to provide the network manager with a synthetic representation of the network situation. To perform this task, visualisation tools can:

- Assist network managers in detecting anomalies and potential threats through an intuitive display of the progression of network traffic.

- Deal easily with highly heterogeneous and noisy data such as the data that is required for network monitoring and ID [44].
- Provide network managers with automated support and motivate their effectiveness by exploiting the ability of the human eye to extrapolate normal traffic patterns and detect anomalies. As stated in [45], "*a picture is worth a thousand packets*" or "*a picture is worth a thousand log entries*" [46].
- Help network managers diagnose performance issues or understand communication patterns between nodes.
- Serve as tools that are complementary to other security mechanisms.

The monitoring task that detects intrusive or anomalous events can be achieved by visualising data at different levels of abstraction: network nodes, intrusion alerts, packet-level data, and so on. In other words, different statistics from various security tools can be visualised, which not only focus on traffic data but also on the network topology.

Visualisation tools rely on human ability to recognize different features and to detect anomalies through graphical devices [42]. Human vision can rapidly locate, discover, identify, and compare objects; all essential tasks in the network monitoring process, considering the overwhelming amount of information and raw traffic data that must be processed [45].

To date, most IDSs have approached ID from a classification standpoint. They perform a 2-class classification of network traffic: normal/anomalous in anomaly-based ID and intrusive/non-intrusive in misuse-based ID. This work proposes visualisation techniques for ID and network monitoring. It is worth emphasizing that the proposal entails the visualisation of network traffic data for detecting intrusions (that is, visualisation for ID) and not the visualisation of IDS alerts or logs (that is visualisation of ID) as others have done [47].

Fig. 2.3, taken from [48], depicts the mutually supporting capabilities of classification (or automated) and visualisation-based IDSs. By combining the classification and visualisation approaches, both the accuracy and completeness of IDSs can be greatly improved. The shaded area shows the increased coverage provided by visualisation-based IDSs.

The visualisation-based approach to ID relies on the following ideas [48]:

- Anomalous situations can be identified by their "visual signature". Visual fingerprints are frequently visible despite the visual noise of background traffic.
- Some stealthy attacks are resistant to detection by classification-based IDSs, but are readily visible using appropriate visualisations.
- Visualisation techniques require little resources and are remarkably resistant to overload caused by high volumes of network traffic.
- The completeness of visualisation-based IDSs is supposedly higher than that of classification-based IDSs when facing 0-day attacks.

Unlike other security tools, IDSs need to be monitored to make the most of their benefits [20]. The huge number of alerts that are usually generated by IDSs (including many false positives and negatives) is a hindrance to permanent (24h.) monitoring, mainly due to economical costs. Visualisation-based IDS can ease this task, by providing an easily understandable snapshot of the status of the network, reducing the time needed for ID.

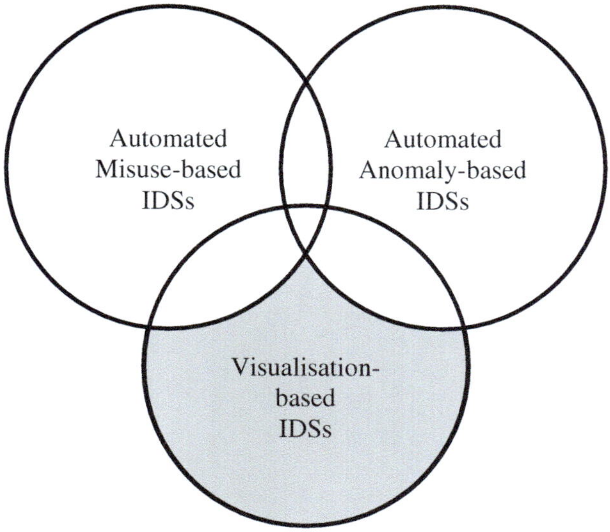

Fig. 2.3 Mutually supporting capabilities of IDSs (from [48]).

The underlying operational assumption of this approach is mainly grounded in the ability to render the high-dimensional traffic data in a consistent yet low-dimensional representation. So, security visualisation tools have to map high-dimensional feature data into a low-dimensional space for presentation. One of the main assumptions of the research presented in this book is that neural projection models will prove themselves to be satisfactory for the purpose of traffic visualisation through dimensionality reduction.

One of the main drawbacks of the visualisation approach to ID is that even being equipped with the "perfect" visualisation technique, security personnel will make mistakes in detecting intrusions. This is a consequence of relying on human abilities as they are affected by a range of factors such as pressure on time, tiredness, boredom, and so on.

Many researchers have previously focused on ID from an information visualisation standpoint. Visualisation techniques are introduced in the next section, while further discussion on their application to network monitoring and ID in the past is provided in Chapter 3.

2.4 Visualisation Techniques

Visualisation techniques have been widely covered in literature; two of the most relevant works being [49, 50]. This section briefly introduces some of the most common sets of these techniques:

- **Scatter plots:** defined as "*the simplest and (arguably most powerful) technique for visualization*" [23], these plots represent bidimensional data as points on a plane, with coordinates that correspond to their values. Due to their simplicity, these types of plots remain one of the most popular and widely-used visual representations for multidimensional data [51].

Two of their main drawbacks are the required low dimensionality of the data to be displayed and the problem of overplotting.

Fig. 2.4 shows a sample scatter plot of the two first dimensions (sepal length and sepal width) of the well-known Iris dataset [52]. Colour is also used to distinguish between the three different classes in the dataset.

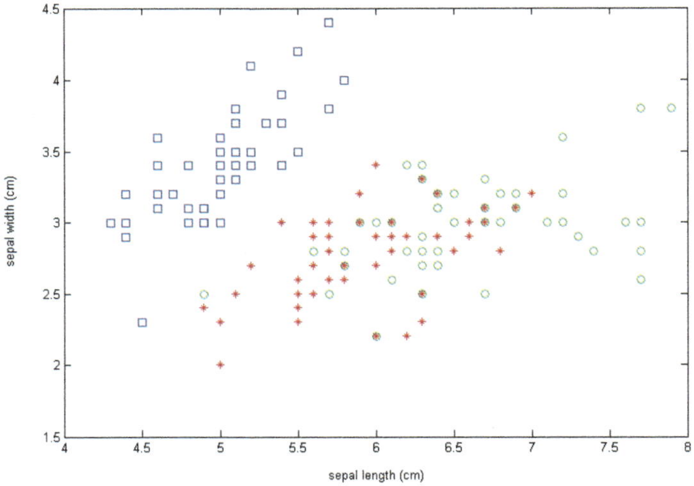

Fig. 2.4 Sample scatter-plot representation of the Iris dataset.

- **Pair plots** (or scatter plot matrixes)**:** these may be defined as extensions of scatter plots, in which each pair of dimensions in the data are plotted on a separate plot. One of their main drawbacks as a visualisation technique is their redundancy and the amount of unused space (due to symmetry). Pair plots can usefully display up to a dozen variables, although the user may have trouble when visually processing so much data.

  Fig. 2.5 shows the pair plot of the well-known Iris dataset, which includes all the dimensions in the dataset (sepal length, sepal width, petal length, and petal width).

- **Parallel coordinate plots:** these are based on placing the coordinate axes in positions that are parallel rather than perpendicular to each other (as in the case of scatter plots). Thus, a multidimensional vector can be visualised straightaway. Their drawbacks relate to problems of overplotting and interpretation.

  Fig. 2.6 shows the parallel coordinate plot of the Iris dataset, which includes all the dimensions in the dataset (sepal length, sepal width, petal length, and petal width).

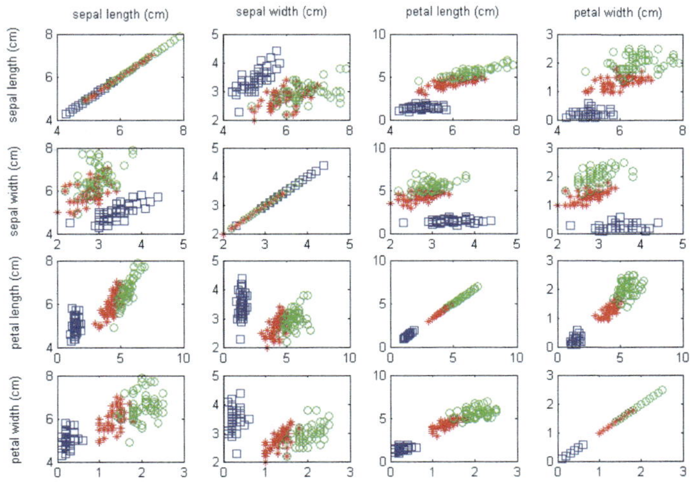

Fig. 2.5 Sample pair-plot representation of the Iris dataset.

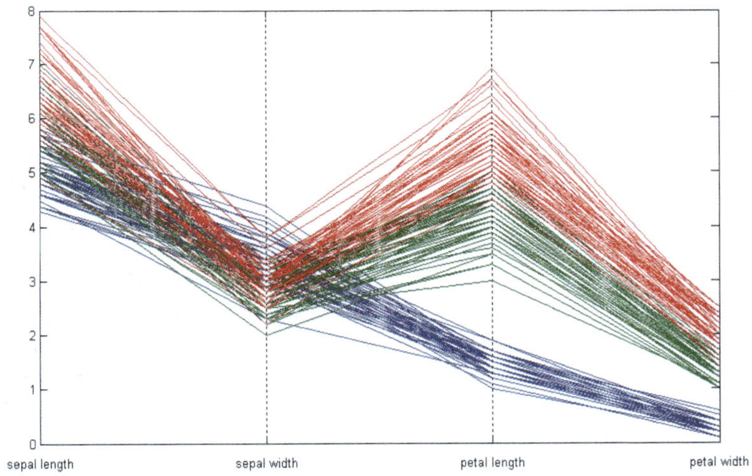

Fig. 2.6 Sample parallel coordinate representation of the Iris dataset.

Parallel coordinate plots can also be depicted in polar coordinates, generating the so called "star plots". In this case, values are plotted on axes that extend from a central point. It is the general shape of each star plot, when compared to other star plots, that is found to provide the most useful insight [45], as can be seen below in Fig. 2.7.

This figure shows a sample multiple-star plot of the eight first instances of the Iris dataset. All the dimensions in the dataset (sepal length, sepal width, petal length, and petal width) are depicted for each instance.

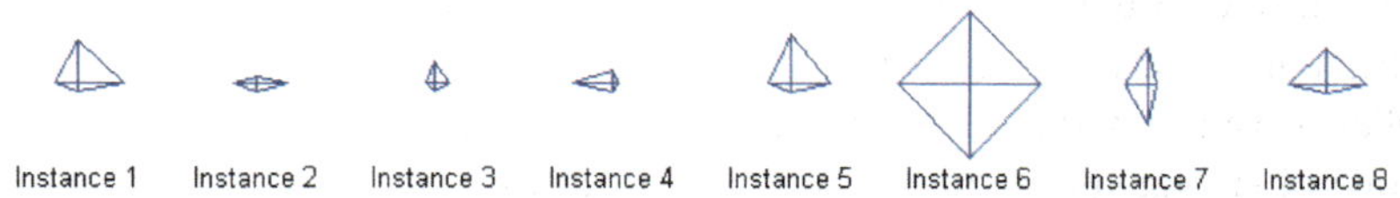

Fig. 2.7 Sample multiple-star plot representation of the Iris dataset.

- **Colour histograms** (also know as data images or matrixes): these enable the visualisation of high-dimensional data. Each dataset is treated as an image in which the columns correspond to observations and the rows to dimensions, or vice versa [23]. The colour of each cell (or pixel) is chosen according to the given column-row value.
- **Charts:** there is a wide range of different charts (line chart, pie chart, bar chart, sector chart, stacked area chart, bubble chart, histograph, etc.) that can summarize multidimensional information in different ways. One of the weaknesses of charts is that they can complicate the task of data interpretation.

The depiction of data through visual techniques is neither an easy nor an immediate task. The difficulty lies in converting raw data into a graphical format that provides useful insight into the visualised dataset [45].

2.5 Artificial Neural Networks

Since its inception, the study of Artificial Neural Networks (ANNs) has been grounded in the awareness that the human brain computes in an entirely different way from the conventional digital computer. The relatively slow operation of human neurons (nerve cells) is overcome by the truly staggering number of neurons in the human brain; each with a massive number of interconnections. Studies have estimated that there are around 10 billion neurons in the human cortex, and 60 trillion connections -synapses- between them, which give one an idea of its highly complex structure.

As well as being highly complex, the brain is also a nonlinear, parallel information-processing system. Neurons are organised within the brain to perform certain computations (e.g. pattern recognition, perception, and motor control) much faster than the fastest digital computer. As an example, it is known that the brain routinely accomplishes perceptual recognition tasks (e.g. recognizing a previously known face embedded in an unknown environment) in the order of 100-200 ms, while a digital computer will take much longer to perform the same task (if indeed it can).

An ANN is a system designed to simulate the way in which the brain performs a particular task or function of interest. The network is usually implemented by using electronic components or simulation software on a digital computer. To achieve high performance, ANNs use a massive interconnection of simple computing cells referred to as "neurons" or "processing units". In such a network, knowledge is obtained through a learning process. This knowledge is stored in the interneuron connection strengths, known as synaptic weights. The procedure for modifying the synaptic weights of the network in an orderly way to reach a desired objective is called the learning algorithm.

2.5.1 Artificial Neuron

As previously stated, a neuron can be defined as the constituent element of an ANN: the neuron is the smallest processing unit within the ANN. They were initially designed to mimic the human neurons as the constituent element of the neural system. A model of an artificial (or synthetic) neuron was initially proposed in [53] and is depicted in Fig. 2.8.

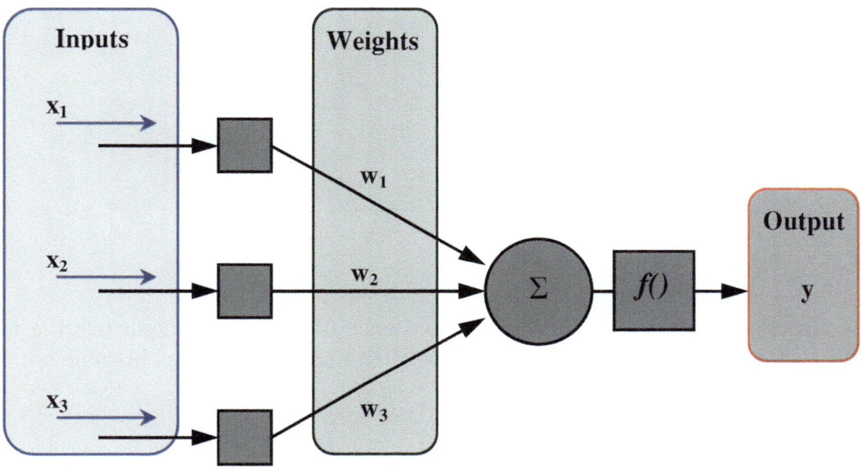

Fig. 2.8 Artificial neuron model.

The network inputs ($\mathbf{x}$) are passed through the synaptic weights ($\mathbf{w}$) and are then summed. Subsequently, the activation function ($f()$) is applied to calculate the output of the neuron ($\mathbf{y}$).

An ANN usually consists of several neurons organised according to a certain network topology. An example of a network topology consisting of an input layer, two hidden layers and an output layer is shown in Fig. 2.9.

The following sections describe the different ways in which the layer connections (weights) of an ANN are modified.

2.5.2 Learning Algorithms

There are three main types of learning algorithms for automated weight setting in ANNs: supervised learning, unsupervised learning, and reinforcement learning.

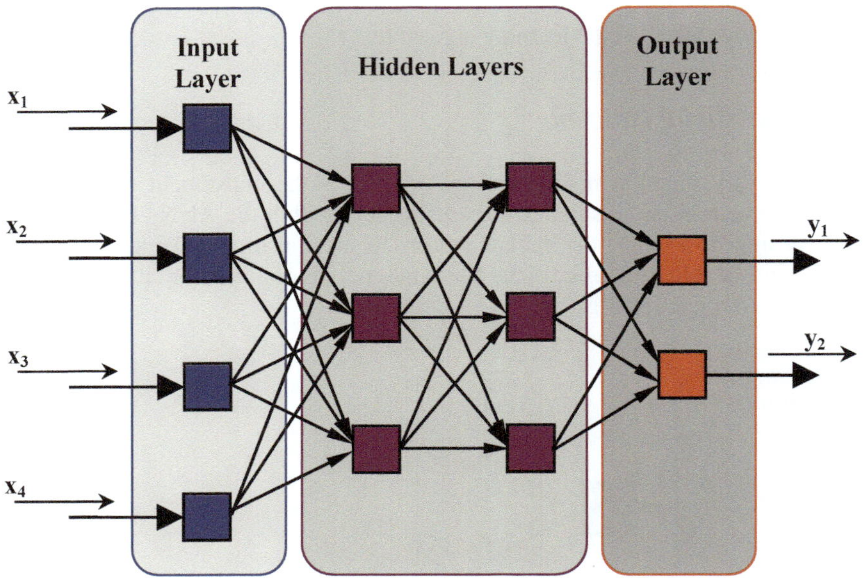

Fig. 2.9 Sample ANN architecture.

Supervised learning relies on the existence of the target output that the network should ideally generate. The elements required in supervised learning are:

- Input of the network: numerical values that are introduced in the network according to the problem setting.
- Internal dynamics of the network: determine the output of the network related to the input.
- Evaluation of the target: generates the "measure" to update the weights.

Typically, ANNs based on supervised learning use error descent to modify their weights.

In contrast, for **unsupervised learning** only the first two elements (input and internal dynamics of the network) are available. No external information is used to check on the weight setting process.

Human beings appear to be able to learn without explicit supervision. Thus, unsupervised learning mimics this aspect of human learning; hence this type of learning tends to use more biologically plausible methods than those using error descent methods. A pertinent example is the local processing at each synapse in these algorithms, which involves no global information passing through the network. So, an unsupervised network must self-organise with respect to its internal parameters, without external prompting. To do so, it must react to some aspect of the input data; typically, redundancy, or clusters. There are, moreover, two major methods of unsupervised learning: Hebbian learning and competitive learning.

Finally, **reinforcement learning** combines the two other types of learning. Although some supervision is provided by telling the network whether the calculated output is right or wrong, there is no evaluation of the target as in the case of supervised learning. Changes in the network weights are only required when the output of the network is wrong.

2.5.3 Hebbian Learning

Hebbian learning is named after Donald Hebb who conjectured:

"*When an axon of a cell A is near enough to excite a cell B and repeatedly or persistently takes part in firing it, some growth process or metabolic change takes place in one or both cells such that A's efficiency as one of the cells firing B, is increased*" [54].

This statement is sometimes expanded [55] into a two-part rule:

1. If two neurons on either side of a synapse are activated simultaneously (i.e. synchronously), then the strength of that synapse must be selectively increased.
2. If two neurons on either side of a synapse are activated asynchronously, then that synapse must be selectively weakened or eliminated.

Considering a basic feed-forward neural network, this would be interpreted in the following way: the weight between an input neuron and an output neuron is greatly strengthened when the activation of the input neuron causes the output neuron to fire strongly. It can be seen that the rule favours the strong: if the weights between inputs and outputs are already large (and so an input will have a strong effect on the outputs), it is likely that the weights will grow.

In mathematical terms, if we consider the simplest feed-forward neural network which has an input layer with associated input vector, $\mathbf{x}$, and an output layer with associated output vector, $\mathbf{y}$, we have the expression:

$$y_i = \sum_j W_{ij} x_j \tag{1}$$

where W_{ij} represents the weight vector between input *j* and output *i*.

The Hebbian learning rule is defined by:

$$\Delta W_{ij} = \eta x_j y_i \tag{2}$$

where η is the learning rate parameter. This means that the weight between each input and output neuron is increased in proportion to the magnitude of the simultaneous firing of these neurons.

The value of $\mathbf{y}$, calculated by the feed forward of the activity, can be substituted in the learning rule to get:

$$\Delta W_{ij} = \eta x_j \sum_k W_{ik} x_k = \eta \sum_k W_{ik} x_k x_j \tag{3}$$

The statistical properties of the learning rule are emphasised in this last expression. It can be seen how the learning rule depends on the correlation between different parts of the inputs data's vector components.

Usually, the final values of the weights are the most interesting part of the network after training. These will enable the prediction of how the network will react to new inputs. However, this is only possible if the weights converge to specific set values.

Most of the neural methods discussed in this chapter are based on Hebbian learning.

2.5.4 Anti-Hebbian Learning

If we correlate the firing of neurons in a neural network, then each contains information about the other. So there is redundancy of information in the firings. Anti-Hebbian learning is used to decorrelate the neuron responses. In that way, more information can be passed through a network when the nodes of the network are all dealing with different data. The goal is to get neurons that respond to different stimuli. The anti-Hebbian learning rule is represented by:

$$\Delta W_{ij} = -\eta y_i y_j \tag{4}$$

This means that if two outputs, y_i and y_j, are highly correlated then the weights between them will grow to a large negative value and each will tend to turn the other off.

It has been shown [56, 57] that weights updated by anti-Hebbian learning will converge toward values such that the corresponding y_i and y_j values are decorrelated. There is no need to limit the growth of such weights as the anti-Hebbian learning process is inherently self-limiting.

2.5.5 Competitive Learning

There are limited resources for learning in the human brain, so gaining one neuron means the loss of another. This fundamental point is imitated in competitive learning. In this kind of learning, the neurons in the output layer of a network compete between themselves to become the one to be fired, as only one single output neuron is active at any one time. This is the biggest difference with respect to Hebbian learning, in which several output neurons may be active simultaneously.

There are three basic elements in a competitive learning rule according to [58]:

- **Set of neurons:** similar except for some randomly distributed synaptic weights. As a consequence, their responses to a given set of input patterns are different.
- **Limit:** imposed on the "strength" of each neuron.
- **Competing mechanism:** permits neurons to compete when responding to a given subset of inputs, such that only one output neuron, or only one neuron per group, is fired at a time. The neuron that wins this competition is called the "winner-takes-all" neuron.

So, each individual neuron learns to specialize on sets of similar patterns, and thereby becomes a feature detector.

2.5.6 Principal Component Analysis

Principal Components Analysis (PCA) is a statistical model introduced in [59] and independently in [60] to describe the variation in a set of multivariate data in terms of a set of uncorrelated variables each of which is a linear combination of the original variables.

Its goal is to derive new variables, in decreasing order of importance, that are linear combinations of the original variables and are uncorrelated with each other. From a geometrical point of view, this goal mainly consists of a rotation of the axes of the original coordinate system to a new set of orthogonal axes that are ordered in terms of the amount of variance of the original data they account for.

Using PCA, it is possible to find a smaller group of underlying variables that describe the data which are called the principal components of the data. So the first few principal components could be used to explain most of the variation in the original data. Note that, even if we are able to characterise the data with a few variables, it does not follow that we will be able to assign an interpretation to these new variables.

A key challenge in the analysis of high dimensional data is to identify the patterns that exist across dimensional boundaries. Such patterns may become visible if a change is made to the basis of the space, however an *a priori* decision as to which basis will reveal most patterns requires prior knowledge of the unknown patterns. PCA offers a potential solution to this problem as it tries to find the orthogonal basis which maximises the data's variance for a given dimensionality of basis. The typical way to do so is to find the direction which accounts for most of the data's variance; this is the first basis vector (the first principal component direction). One can then find the direction that accounts for most of the remaining variance (the second basis vector) and so on. If one then projects the data onto the principal component directions, a dimensionality reduction is performed. It is optimized by considering the retention of variance (or information) in the data. The basis vectors of this new co-ordinate system are the eigenvectors of the covariance matrix of the dataset and the variance on these co-ordinates are the corresponding eigenvalues. The optimal projection given by PCA from an N -dimensional to an M -dimensional space is the subspace spanned by the M eigenvectors with the largest eigenvalues.

PCA can also be thought of as a data compression technique, where there is minimum information loss in the data in terms of least mean square error. As a result, it is often used as a pre-processing method in order to simplify a high dimensional space prior to its analysis.

According to [61], it is possible to describe PCA as a mapping of vectors $\mathbf{x}^{\mathbf{d}}$ in an N -dimensional input space $(x_1,...,x_N)$ onto vectors $\mathbf{y}^{\mathbf{d}}$ in an

M -dimensional output space $(y_1,...,y_M)$, where $M \le N$. $\mathbf{x}$ may be represented as a linear combination of a set of N orthonormal vectors W_i :

$$\mathbf{x} = \sum_{i=1}^{N} y_i W_i \tag{5}$$

Vectors W_i satisfy the orthonormality relation:

$$W_i^t W_j = \delta_{ij} \tag{6}$$

where δ_{ij} is the Kronecker delta.

Making use of equation (5), the coefficients y_i may be given by

$$y_i = W_i^T \mathbf{x} \tag{7}$$

which can be regarded as a simple rotation of the co-ordinate system from the original $\mathbf{x}$ values to a new set of co-ordinates given by the $\mathbf{y}$ values. If only one subset $M < N$ of the basis vectors, W_i, is retained so that only M coefficients y_i are used, and having replaced the remaining coefficients by constants b_i, then each $\mathbf{x}$ vector may be approximated by the following expression:

$$\tilde{\mathbf{x}} = \sum_{i=1}^{M} y_i W_i + \sum_{i=M+1}^{N} b_i W_i \tag{8}$$

Consider the whole dataset of D vectors, $\mathbf{x^d}$ where $d = 1,..., D$.

It is necessary to make the best choice of W_i and b_i so that the y_i values obtained by equation (7) give the best approximation of equation (8) over the whole dataset. Vector $\mathbf{x^d}$ has an error due to dimensionality reduction:

$$\mathbf{x^d} - \tilde{\mathbf{x}}^\mathbf{d} = \sum_{i=M+1}^{N} \left(y_i^d - b_i\right) W_i \tag{9}$$

The best approximation can be defined as that which minimises the sum of the squares of the errors over the whole dataset

$$E_M = \frac{1}{2} \sum_{\mathbf{d}=1}^{\mathbf{D}} \left\| \mathbf{x^d} - \tilde{\mathbf{x}}^\mathbf{d} \right\|^2 = \frac{1}{2} \sum_{d=1}^{D} \sum_{i=M+1}^{N} \left(y_i^d - b_i\right)^2 \tag{10}$$

If we calculate the derivative of E_M with respect to b_i and set it to zero then:

$$b_i = \frac{1}{D} \sum_{d=1}^{D} y_i^d = W_i^T \overline{\mathbf{x}} \quad \forall i \;\; M+1,..., N \tag{11}$$

where the sample mean vector $\bar{\mathbf{x}}$ is defined as:

$$\bar{\mathbf{x}} = \frac{1}{D}\sum_{d=1}^{D}\mathbf{x}^{\mathbf{d}} \tag{12}$$

Now the sum of square errors can be written as:

$$E_M = \frac{1}{2}\sum_{i=M+1}^{N}\sum_{d=1}^{D}\left\{W_i^T\left(\mathbf{x}^{\mathbf{d}} - \bar{\mathbf{x}}\right)\right\}^2 = \frac{1}{2}\sum_{i=M+1}^{d}W_i^T\mathbf{\Sigma}W_i \tag{13}$$

where Σ is the sample covariance matrix of the set of vectors $\left\{\mathbf{x}^{\mathbf{d}}\right\}$ and is given by:

$$\Sigma = \frac{1}{D}\sum_{d}\left(\mathbf{x}^{\mathbf{d}} - \bar{\mathbf{x}}\right)\left(\mathbf{x}^{\mathbf{d}} - \bar{\mathbf{x}}\right)^T \tag{14}$$

Then if we minimise E_M with respect to the choice of W_i it can be shown [61] that the minimum occurs when the basis vectors satisfy

$$\Sigma W_i = \lambda_i W_i \tag{15}$$

Hence, those W_i are the eigenvectors of the covariance matrix. As is assumed, the eigenvectors can be proved to be orthogonal, if the covariance matrix is real and symmetric. If we substitute equation (15) into equation (13) and make use of the orthonormality relation in equation (6), then the value of the error criterion at the minimum may be represented in the form:

$$E_M = \frac{1}{2}\sum_{i=M+1}^{N}\lambda_i \tag{16}$$

Subsequently, the minimum error is obtained by choosing the smallest eigenvalues (N - M), and their corresponding eigenvectors, as the ones to be discarded. We usually refer to the y's as the principal components.

PCA can be performed by means of ANNs or connectionist models such as [62, 63, 64, 65, 66]. It should be noted that even if we are able to characterize the data with a few variables, it does not follow that an interpretation will ensue.

2.5.7 Oja's Weighted Subspace Algorithm

A model of Hebbian learning with weight decay was introduced in [64], which not only stops the weights growing without bound, but also causes them to converge to the principal components of the input data. This algorithm was extended in [67] to identify the actual principal components using the Weighted Subspace

Algorithm. The authors recognized the importance of introducing asymmetry into the weight decay process in order to force weights to converge to the principal components. The resulting model is defined by:

$$y_i = \sum_{i=1}^{N} W_{ij} x_j \quad (17)$$

where a Hebb-type rule with decay modifies the weights according to:

$$\Delta W_{ij} = \eta y_i \left(x_j - \theta_i \sum_{k=1}^{M} y_k W_{kj} \right) \quad (18)$$

Ensuring that $\theta_1 < \theta_2 < \theta_3 < \ldots$ allows the neuron whose weight decays in proportion to θ_1 to learn the principal values of the correlation in the input data. As a consequence, this neuron will maximally respond to directions parallel to the first principal component. The second output can not compete with the first, but it is in a stronger position to identify the second principal component and so on for all of the outputs in the network.

2.5.8 Negative Feedback Network

Feedback exists in a network whenever the output of an element in the network partially influences the input applied to that particular element. This mechanism is used in the case of the Negative Feedback Network (NFN) [68] to maintain the equilibrium of the weight vectors. To define this model, consider an N -dimensional input vector ($\mathbf{x}$), and an M -dimensional output vector ($\mathbf{y}$), with W_{ij} being the weight linking input j to output i and let η be the learning rate.

Initially, there is no activation at all in the network. The input data is fed forward throughout the weights from the input neurons (the $\mathbf{x}$ -values) to the output neurons (the $\mathbf{y}$ -values) where a linear summation is performed to give the activation of the output neuron. This is expressed as:

$$y_i = \sum_{j=1}^{N} W_{ij} x_j, \forall i \quad (19)$$

The activation is fed back through the same weights and subtracted from the inputs. It is expressed in [69] as follows:

$$x'_j = x_j - \sum_{i=1}^{M} W_{ij} y_i, \ \forall j \quad (20)$$

The term " $\mathbf{e}$ " may be used instead of " $\mathbf{x}'$ ", allowing equation (20) to be rewritten as:

$$e_j = x_j - \sum_{i=1}^{M} W_{ij} y_i, \ \forall j \tag{21}$$

Subsequently, simple Hebbian learning is performed between input and outputs:

$$\Delta W_{ij} = \eta e_j y_i \tag{22}$$

The stabilization of the learning in the network is the effect of the negative feedback. As a consequence of that, it is not necessary to normalise or clip the weights to ensure convergence to a stable solution.

Note that this algorithm is clearly equivalent to Oja's Subspace Algorithm [62] as:

$$\Delta W_{ij} = \eta e_j y_i = \eta \left(x_j - \sum_k W_{kj} y_k \right) y_i \tag{23}$$

The NFN is then capable of finding the principal components of the input data [70] in an equivalent way to Oja's Subspace algorithm [62]. Instead of finding the principal components, the weights of such a network will identify a basis of the subspace spanned by these components. We may emphasize that the NFN uses simple Hebbian learning that enables the weights to converge in order to extract the maximum information content from the input data.

Writing the algorithm in this way allowed different versions and algorithms such as the Maximum Likelihood Hebbian Learning rule [71] to be devised. It is based on an explicit view of the residual $(\mathbf{x} - W\mathbf{y})$ which is never independently calculated using Oja's learning rule.

2.5.9 Nonlinear Principal Component Analysis

Non-linear PCA is a fairly popular nonlinear method [72, 73, 74] which a number of authors [130, 165, 166, 253] have quite successfully applied to Independent Component Analysis (ICA) [75]. ICA is a special case in the world of independence seeking networks in which linear mixtures of independent signals at the inputs of a network are completely separated from the outputs. The dimensionality of the inputs and the dimensionality of the outputs are generally the same. [72] has introduced the following nonlinear extension to Oja's Subspace Algorithm [62]:

$$\Delta W_{ij} = \eta \left(x_j f(y_i) - f(y_i) \sum_k W_{kj} f(y_k) \right) \tag{24}$$

which can be derived as an approximation of the best nonlinear compression of the data in the following way.

Starting with the cost function:

$$\mathbf{J}(W)=1^T E\left\{\left(\mathbf{x}-Wf\left(W^T\mathbf{x}\right)\right)^2\right\} \tag{25}$$

the aim is to minimise the sum of the squared representation errors for the $\mathbf{x}$ vector. By taking the instantaneous gradient and implementing a stochastic gradient descent, we may derive the following weight update rule:

$$\Delta W \propto \frac{-\delta J(W)}{\delta W} \propto \eta\left(\mathbf{x}\mathbf{e}^{\mathrm{T}}Wf'\left(\mathbf{x}^{\mathrm{T}}W\right)+\mathbf{e}f\left(\mathbf{x}^{\mathrm{T}}W\right)\right) \tag{26}$$

where the error vector is $\mathbf{e}=\left(\mathbf{x}-Wf\left(W^T\mathbf{x}\right)\right)$, and $f'\left(\mathbf{x}^{\mathrm{T}}W\right)$ is the element-wise derivative of $f\left(\mathbf{x}^{\mathrm{T}}W\right)$ with respect to W. In [72], it is argued that the term $\mathbf{x}\mathbf{e}^{\mathrm{T}}Wf'\left(\mathbf{x}^{\mathrm{T}}W\right)$ has a negligible effect on the learning compared to the second term, $\left(\mathbf{e}f\left(\mathbf{x}^{\mathrm{T}}W\right)\right)$, and may therefore be disregarded, yielding:

$$\Delta W=\eta\left(\left(\mathbf{x}-Wf\left(W^T\mathbf{x}\right)\right)f\left(\mathbf{x}^{\mathrm{T}}W\right)\right) \tag{27}$$

This rule has been used to extract a higher order structure than the PCA network, such as the independent components of data [72, 76].

2.5.10 Exploratory Projection Pursuit

Exploratory Projection Pursuit (EPP) [77] is a statistical method aimed at solving the difficult problem of identifying structure in high dimensional data. It is done by projecting the data onto a low dimensional subspace in which structure may be visually identified. As not all projections will reveal the data's structure equally well, it defines an index that measures the extent to which a given projection is "interesting", and then represents the data in terms of projections that maximise that index.

Then, the first step for EPP is to define which indices represent interesting directions. Concerning projections, "interestingness" is usually defined with respect to the fact that most projections of high-dimensional data give almost Gaussian distributions [78]. Thus, those directions which reveal data-projections that are as far from the Gaussian as possible should be found in order to identify the most "interesting" features of the data. Transforming the data into a zero mean and identity covariance matrix enables a network to respond solely to higher order statistics, which can be used to look for interesting structure in the data.

There are two simple measures of deviation from a Gaussian distribution based on the higher order moments of the distribution: skewness and kurtosis. Skewness is based on the normalised third moment and measures the deviation of the distribution from bilateral symmetry. Kurtosis is based on the normalised fourth moment and measures the heaviness of the tails of a distribution. A bimodal distribution will often have a negative kurtosis and therefore negative kurtosis can signal that a particular distribution shows evidence of clustering.

A Gaussian distribution with mean a and variance x is no more or less interesting than a Gaussian distribution with mean b and variance y. Furthermore, this second order structure can obscure a higher order and more interesting structure. As a consequence, such information is removed from the data in a process known as "sphering": raw data is translated till its mean equals zero, projected onto the principal component directions, and multiplied by the inverse of the square root of its eigenvalue, to give data with mean zero and unit variance in all directions.

2.5.11 The Exploratory Projection Pursuit Network

The EPP neural network [79] is essentially a nonlinear modification of Oja's subspace algorithm [62]. The network can be described by the following set of equations:

$$s_i = \sum_{j=1}^{N} W_{ij} x_j \tag{28}$$

$$e_j \leftarrow x_j - \sum_{k=1}^{M} W_{kj} s_k \tag{29}$$

$$r_i = f(s_i) \tag{30}$$

$$\Delta W_{ij} = \eta r_i e_j \tag{31}$$

where x_j is the sphered activation of the jth input, s_i is the activation of the ith output neuron, W_{ij} is the weight between the former and the latter, and r_i is the value of the function $f()$ on the ith output neuron.

Initially there is no activation in the network. The input data is fed forward via the weights, W, to the output neurons where a simple summation is performed. The activations of the output neurons are fed back via the same weights to the input neurons as inhibition and are therefore subtracted. Then a (nonlinear) function of the weights is calculated and used to update the weights by applying the simple Hebbian learning rule.

It was shown in [72] that the use of a nonlinear function $f()$ in the above equations creates an algorithm to find those values of W that maximize that function, subject to the constraint that W is an orthogonal matrix. This idea was applied in [79] to the previously described network in the context of the network performing EPP. Thus, the function $f(s) \approx s^3$ is applied to the algorithm if we wish to find the direction which maximises the kurtosis of the distribution that is measured by s^4; similarly, the function $f(s) \approx s^2$ is applied to find the direction with maximum skewness.

2.5.12 Cooperative Maximum Likelihood Hebbian Learning

Cooperative Maximum Likelihood Hebbian Learning (CMLHL) is based on the EPP neural model called Maximum Likelihood Hebbian Learning (MLHL) [71, 80]. The main difference between these two models is that CMLHL includes lateral connections [81, 82] derived from the Rectified Gaussian Distribution (RGD) [83]. The RGD is a modification of the standard Gaussian distribution in which the variables are constrained to be non-negative, enabling the use of non-convex energy functions. In a more precise way, CMLHL includes lateral connections based on the mode of the cooperative distribution that is closely spaced along a nonlinear continuous manifold. By including these lateral connections, the resulting network can find the independent factors of a dataset in a way that captures some type of global ordering in the dataset.

Considering an N-dimensional input vector $\mathbf{x}$, an N-dimensional output vector $\mathbf{y}$ and with W_{ij} being the weight (linking input j^{th} to output i^{th}), CMLHL can be expressed as:

Feed-forward step:

$$y_i = \sum_{j=1}^{N} W_{ij} x_j, \forall i \tag{32}$$

Lateral activation passing:

$$y_i(t+1) = [y_i(t) + \tau(b - Ay)]^+ \tag{33}$$

Feedback step:

$$e_j = x_j - \sum_{i=1}^{M} W_{ij} y_i, \forall j \tag{34}$$

Weight change:

$$\Delta W_{ij} = \eta . y_i . sign(e_j) | e_j |^p \tag{35}$$

where η is the learning rate, τ is the "strength" of the lateral connections, b the bias parameter, and p a parameter related to the energy function [71, 80, 81].

A is a symmetric matrix used to modify the response to the data whose effect is based on the relation between the distances among the output neurons. It is based on the cooperative distribution, but to speed learning up, it can be simplified to:

$$A(i, j) = \delta_{ij} - \cos(2\pi(i - j)/M) \tag{36}$$

where δ_{ij} is the Kronecker delta.

The application of CMLHL, initially in the field of artificial vision [81, 82], and subsequently to other interesting topics [84, 85, 86], has proven that this model can successfully perform data visualisation.

2.5.13 Self-Organizing Map

The Self-Organizing Map (SOM) [87] was developed as a visualisation tool for representing high dimensional data on a low dimensional display. It is also based on the use of unsupervised learning, although it is a topology preserving mapping model rather than a projection architecture. A SOM, composed of a discrete array of L nodes arranged on an N -dimensional lattice, maps these nodes into a D -dimensional data space while preserving their ordering. The dimensionality of the lattice (N) is normally smaller than that of the data, in order to perform the dimensionality reduction. The SOM can be viewed as a nonlinear extension of PCA, where the map manifold is a globally nonlinear representation of the training data [88].

Typically, the array of nodes is one or two-dimensional, with all nodes connected to the N inputs by an N -dimensional weight vector. The self-organization process is commonly implemented as an iterative on-line algorithm, although a batch version also exists. An input vector is presented to the network and a winning node, whose weight vector W_c is the closest (in terms of Euclidean distance) to the input, is chosen:

$$c = \arg\min_i (\|\mathbf{x} - W_i\|) \tag{37}$$

So the SOM is a vector quantiser, and data vectors are quantised to the reference vector in the map that is closest to the input vector. The weights of the winning node and the nodes close to it are then updated to move closer to the input vector. There is also a learning rate parameter (η) that usually decreases as the training process progresses. The weight update rule is defined as:

$$\Delta W_i = \eta h_{ci} [\mathbf{x} - W_i], \forall i \in N^{(c)} \tag{38}$$

When this algorithm is sufficiently iterated, the map self-organises to produce a topology-preserving mapping of the lattice of weight vectors to the input space based on the statistics of the training data.

This connectionist model is applied in this work for comparative purposes, as it is one of the most widely used unsupervised neural models for visualising structure in high-dimensional datasets. The SOM has aroused a great deal of interest in the neural network community. To date, the number of scientific papers written on the SOM is in excess of 5,000 [89]. SOM applications cover robotics, speech recognition, data compression, processing optic patterns and organisation of text documents.

2.5.14 Curvilinear Component Analysis

Curvilinear Component Analysis (CCA) [90] is a nonlinear dimensionality reduction method. It was developed as an improvement on the SOM. It tries to circumvent the limitations inherent in some previous linear models such as PCA. Its output is not a fixed lattice but a continuous space able to take the shape of the submanifold in the dataset (input space).

The principle of CCA is a self-organised neural network performing two tasks: a vector quantization of the submanifold and a nonlinear projection of these quantising vectors toward an output space, providing a revealing view of the way in which the submanifold unfolds. Quantization and nonlinear mapping are separately performed by two layers of connections.

In the vector quantization step, the input vectors are forced to become prototypes of the distribution by using competitive learning and the regularization method [91] of vector quantization. Thus, this step, which is intended to reveal the submanifold of the distribution, regularly quantizes the space covered by the data, regardless of the density. Euclidean distances between these input vectors are considered, as the output layer has to build a nonlinear mapping of the input vectors.

A perfect matching is not possible at all scales when the manifold is "unfolding", so a weighting function is introduced, yielding the quadratic cost function:

$$E = \frac{1}{2}\sum_{i}\sum_{j \neq i}\left(X_{ij} - Y_{ij}\right)^2 F\left(Y_{ij}, \lambda_y\right) \tag{39}$$

As regards its goal, the projection part of CCA is similar to other nonlinear mapping methods, in that it minimizes a cost function based on interpoint distances in both input and output spaces. Instead of moving one of the output vectors (y_i) according to the sum of the influences of every other y_j (as would be the case for a stochastic gradient descent), CCA proposes pinning down one of the output vectors (y_i) "temporarily", and moving all the other y_j around, disregarding any interactions between them. Accordingly, the proposed "learning" rule can be expressed as:

$$\Delta y_j = \alpha(t) F\left(Y_{ij}, \lambda_y\right)\left(X_{ij} - Y_{ij}\right)\frac{y_j - y_i}{Y_{ij}} \quad \forall j \neq i \tag{40}$$

The main advantages of CCA, in comparison to other methods such as the stochastic gradient descent or the steepest gradient descent are:

1. The proposed rule (Δy) is much lighter than a stochastic gradient from a computational standpoint.
2. The average of the output vector updates is proportional to the opposite of the gradient of the cost function (E). On the other hand, it can temporarily produce increases in E , which eventually allows the algorithm to escape from local minima of E . It has been shown [90] that it implies a lower final cost in comparison with gradient methods.

CCA is able to perform dimensionality reduction and represent the intrinsic structure of given input data without any previous knowledge about the distribution of the analysed dataset. Compared with other previous projection algorithms, the CCA method is more general, reliable, and faster at capturing input data structure.

2.6 Agents and Multiagent Systems

Modern-day demand for complex and yet effective systems have led to the development of new computing paradigms, amongst which figure Agents and MAS.

The concept of an agent was originally conceived in the field of AI, and subsequently evolved as a computational entity in the field of software engineering. From the software standpoint, it is seen as an evolution to overcome the limitations inherent to the object oriented methodologies. There is as yet no strict definition of what constitutes an agent [92]. In a general AI context, a rational agent was defined in [93] as anything that perceives its environment through sensors and acts upon that environment through effectors. More specifically, a software agent has been defined as a system with the capacity to adapt, which has mechanisms that allow it to decide what to do (according to its objectives) [94]. Additionally, from a distributed AI standpoint, it was defined [95] as a physical or virtual entity:

- Which is capable of acting in an environment.
- Which can communicate directly with other agents.
- Which is driven by a set of tendencies (in the form of individual objectives or of a satisfaction/survival function which it tries to optimize).
- Which possesses resources of its own.
- Which is capable of perceiving its environment (but to a limited extent).
- Which has only a partial representation of its environment (and perhaps none at all).
- Which possesses skills and can offer services.
- Which may be able to reproduce itself.
- Whose behaviour tends towards satisfying its objectives, taking account of the resources and skills available to it and depending on its perception, its representation and the communications it receives.

Although there is no consensus over what an object requires in order to be an agent [96], it is widely accepted that the following properties define agents:

- **Autonomy:** this is the key feature of agents. It refers to the ability of reacting without external prompting; that is, deciding on one's own.
- **Social skills:** agents must interact with other surrounding agents located in the same environment. They are also referred as "heterosocial" agents, i.e., "*agents who can socialize with agents of different types*" [96]. Hence, communication skills are also required to interact with other agents and (possibly) humans.
- **Action:** agents act on one another and on their environment. Hence, actions may change the state of the environment and the beliefs/states of other agents. Some other features concerning activity may be considered, such as reactive/proactive behaviour.
- **Intelligence, knowledge, and rationality:** some authors claim that agents must exhibit at least a little intelligence. Learning abilities can be also required from agents in order for them to adapt to their environment. Furthermore, agents may be conscious about their own beliefs and states.
- **Mobility:** there is some controversy over whether it is a necessary characteristic of an agent to be able to move from one location to another.

Taking these issues into account, a MAS may be defined as one "that contains an environment, objects and agents (the agents being the only ones to act), relations between all the entities, a set of operations that can be performed by the entities and the changes of the universe in time and due to these actions" [95]. From the standpoint of distributed problem solving [97], a MAS can be defined as a loosely coupled network of problem solvers that work together to solve problems that are beyond the individual capabilities or knowledge of each problem solver. According to [98], the characteristics of MASs are:

- Each agent has incomplete information, or capabilities for solving the problem, thus each agent has a limited viewpoint.
- There is no global system control.
- Data is decentralized.
- Computation is asynchronous.

As a consequence, agents in a MAS are driven by their own objectives as there is no global control unit. They take the initiative according to their objectives and dynamically decide what to do or what tasks other agents have to perform.

Agents and MASs have been widely used over recent years, but have not always been the most appropriate solution. According to [99], the problem to be solved has a number of features which point to the appropriateness of an agent-based solution:

- The environment is open, or at least highly dynamic, uncertain, or complex.
- Agents are a natural metaphor. Many environments are naturally modelled as societies of agents, either cooperating with each other to solve complex problems, or else competing with one-another.

- Distribution of data, control, or expertise, which implies that a centralised solution is at best extremely difficult or at worst impossible.
- Legacy systems, which refer to technologically obsolete software which is nevertheless essential to the functioning of an organisation. Such software cannot generally be discarded (because of the short-term cost of rewriting) and it is often required to interact with other software components. One solution to this problem is to wrap the legacy components, providing them with an "agent layer" functionality.

2.6.1 Agent Taxonomy

Several different classifications of agents have been proposed in the literature [92, 100, 101]. The most widely accepted classifications are grouped under the following headings:

- **Interaction with the user**
  - Interface agents: users can interact with the system by means of commands.
  - Autonomous agents: there is some interaction with the user but it is the agent who decides whether actions must be taken according to changes in the user behaviour.
- **Mobility**
  - Static agents: these agents are placed in a system or network and can not be run in a different environment.
  - Mobile agents: these agents can move from one platform or host to another. They can autonomously decide when and where to move.
- **Inner complexity** [93]
  - Simple reflex agents: these agents, also called stimulus-response agents, react to immediate stimuli. They have no memory at all and their percept sequence is just the immediate environment sensed by the agent's sensors.
  - Reflex agent with state: these agents have a memory of past states and can remember earlier experiences. Memories can be combined with current information from sensors to produce a more sophisticated response to the environment.
  - Goal-based agents: these agents can plan ahead before taking an action in their environment. They can use various search methods of different complexities to search a state space of potential future environments.
  - Utility-based agents: utility is an economic term (from micro economics) related to the concepts of happiness and personal preferences. These agents try to optimize their utility levels. Thus, it is said that these agents are more similar to human beings.

2.6.2 Agent Architecture

Before building a software system under the MAS paradigm, some issues must be analysed. One of the most important issues is to decide which agent architecture can support the required features. Three main agent architectures have been defined [102]:

- **Symbolic:** traditionally, in the AI field, agents have been viewed as a type of knowledge-based system, often called deliberative systems. This architecture is strongly related to the symbolic AI paradigm, which focuses on the following issues: the knowledge that needs to be represented, the way in which it will be represented, the reasoning mechanisms that are to be used, and so on. These kinds of agents, which use highly complex communication mechanisms, are defined by an initial state and a set of plans that have to be met in order to achieve a particular goal.
- **Reactive:** due to criticism of the architectures described above, the concept of a reactive agent was proposed. This entails a set of task accomplishing behaviours, arranged into a "subsumption" hierarchy. Each task behaviour is implemented as a simple finite-state machine, which directly maps sensor input to effector output. As there is no reasoning process, the required communication mechanisms are simple.
- **Hybrid:** an architecture based on a combination of the two previous ones. Belief-desire-intention (BDI) is the most usual and widely used architecture of this type, which typically contains four key data structures:
  - Beliefs: correspond to information the agent has about the world, which may be incomplete or incorrect.
  - Desires: intuitively correspond to the tasks allocated to it.
  - Intentions: correspond to desires that the agent has committed to achieving. In most cases, agents will not be able to achieve all their desires, even if those desires are consistent.
  - Plan library: is a set of plans specifying courses of action that may be followed by an agent in order to achieve its intentions.

  Additionally, a further component called the interpreter, is responsible for updating beliefs from observations made of the world, generating new desires on the basis of these new beliefs, and selecting from among the set of currently active desires some subset to act as intentions. The interpreter must also select an action to perform on the basis of the agent's current intentions and procedural knowledge.

2.7 Case-Based Reasoning

Case-Based Reasoning (CBR) is a general problem solving paradigm. Unlike other major AI approaches, it is able to utilize the specific knowledge of concrete problems (modelled as cases) that have previously been experienced. On the other hand, CBR is an approach to incremental learning, as a new experience is retained each time a new problem is solved.

CBR is founded on the idea that similar problems have similar solutions [103]. To model this, CBR defines the concept of case, that can be seen as a past experience described by the 3-tuple <Problem, Solution, Results>. A case is then composed of a problem description (initial state), the solution applied to solve the problem (the actions executed in order to achieve the objectives), and the result obtained after applying the solution (the final state and the evaluation of the actions executed). As a consequence, a case base (memory) must be maintained to solve new problems. When a new problem is presented, the system executes a CBR cycle, composed of four sequential stages as shown in Fig. 2.10. These stages are processes that can comprise different steps as described in [104].

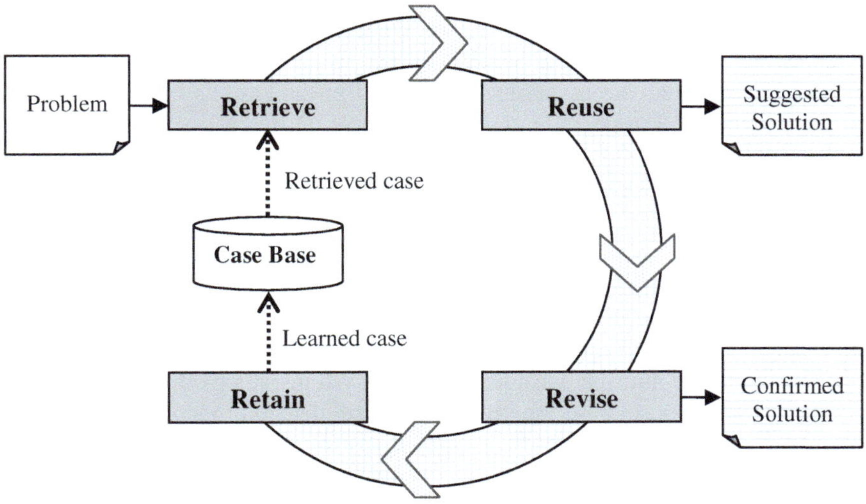

Fig. 2.10 The Case-Based Reasoning cycle.

The four stages in the CBR cycle can be defined as:

- **Retrieve:** once a new problem is defined as a case, those cases with the most similar problem description to the current problem are recovered from the case base. A considerable amount of research has focused on how similarity can be measured under such settings. Two main ideas to assess similarity have been followed [105] based on:
  - Surface features: those that are provided as part of the case description and that are typically represented using attribute-value pairs.
  - Derived features: in some cases, it is necessary to derive new features from the case description. Retrieval therefore requires an assessment of structural similarity of cases, which retrieves more relevant cases.

Although similarity is the most common criteria for retrieving cases, some other features have been considered [105], such as: how effectively the solution space is covered by the retrieved cases, how easily their solutions can be adapted to solve the target problem, or how easily the proposed solution can be explained.

- **Reuse:** the solutions corresponding to the similar cases retrieved in the previous stage are reused to build up a new solution. In classification problems, it might be enough to provide the class of the retrieved case as the solution to the new problem. In more complex settings, the retrieved solution or set of solutions need to be adapted to the new problem. Two main ideas for adaptation have been proposed [106]:
  - Substitution: some part(s) of the retrieved solution are reinstantiated.
  - Transformation: the structure of the solution is altered.
- **Revise:** the newly generated solution is checked. If changes are needed, then adaptation can be performed at this stage.
- **Retain:** the new case (experience) is stored in the case base for future reuse. That is, the CBR system "learns" the new case. Different issues about the new problem can be stored in the case base. The simplest approaches just store the solution while others consider a deeper representation of the problem solving process [105].

In addition to these basic stages, some others have been proposed. That is the case of the review and restore stages, proposed for the maintenance of CBR systems [107]. The review stage monitors the quality of system knowledge while the restore stage selects and applies maintenance operations. Other issues concerning CBR maintenance must be considered in such systems: removal and/or adaptation of cases from the case base, optimizing the size, detecting inconsistencies in the case base, and so on. Maintenance policies can be defined for CBR systems to detail all these issues.

A wide variety of methods are applied in each of the previously described stages: a case can be defined as one problem or a set of problems. There are different ways of indexing cases, and the matching of cases and the adaptation of solutions may also vary. The CBR cycle can be complemented with other problem solving methods. It is especially useful when a similar case can be recovered from the case base due to different reasons. Then, the systems can generate the suggested solution by means of a reasoning method other than CBR.

CBR has been applied to a broad range of systems, that can be classified in a more precise way [104]:

- **Exemplar-based reasoning:** solving a problem is a classification task; in other words, finding the right class for the unclassified exemplar. Thus, the class of the most similar stored case becomes the solution to the classification problem.
- **Instance-based reasoning:** this is a specialization of the previous class into a highly syntactic CBR-approach, as opposed to more knowledge-intensive exemplar-based approaches.

- **Memory-based reasoning:** under this approach, a collection of cases is seen as a "large" memory. Consequently, reasoning is a process of accessing and searching in this memory.
- **Case-based reasoning:** this is a generic term for all these systems, although typical case-based reasoning differs from the approaches described above. Each case has a complex inner structure. Another feature of this approach is that the retrieved solution can be modified, or adapted for its application to the new problem.
- **Analogy-based reasoning:** the differentiating point of this approach is that it solves new problems based on past cases from a different domain. Unlike previous approaches, cases from different problem domains are considered; hence the complexity of reusing past solutions increases.

Chapter 3
Previous Work on NID

Since its inception in the 1980s, IDSs have evolved from monolithic batch-oriented systems to distributed real-time networks of components [40]. As ID has been an active area of research over thirty years or so, since the initial works in this field [28, 29], a vast array of approaches and techniques have been used to build IDSs. As mentioned in the previous chapter, this work proposes a visualisation-based NIDS that applies several AI paradigms from a hybrid system perspective. Thus, this chapter covers state-of-the-art network-based ID. As a visualisation agent-based IDS is proposed in this book, previous work on visualisation and agent - based IDSs is also revised in this chapter. Commercial IDSs are not included in this chapter due to their proliferation on the marketplace [31] over recent years and because manufacturers provide little information on the techniques they implement.

ID has been approached from several different standpoints up to the present day; many different forms of AI (such as Genetic Programming, Data Mining, Expert Systems, Fuzzy Logic, or ANN among others), statistical, and signature verification techniques have been applied. Most of these works have focused on ID from a pattern classification perspective (normal/anomalous in AIDSs and intrusive/non-intrusive in MIDSs). These classification-based IDSs can generate different alarms when an anomalous situation (supposedly an intrusion) is detected, but they can not provide a general overview of what is happening inside a network. To overcome this problem and help security personnel in dealing with the huge amount of ID alarms, visualisation techniques have been also applied to the ID challenge.

Numerous works have also been published that survey the wide range of techniques and data applied to ID [31, 36, 89, 103, 110, 118, 127, 163, 198, 227, 238, 240, 259, 321].

3.1 Overview of Techniques for NID

Initially, the most widely deployed, commercially available IDSs were based on signature analysis [108]. Attack signatures were defined by human experts and then key features extracted from different audit streams were compared to the

Á. Herrero and E. Corchado: Mobile Hybrid Intrusion Detection, SCI 334, pp. 41–70.
springerlink.com

previously defined signatures in order to identify intrusions. SNORT [109] is an open source network-based IDS employing signatures, which is used for performing real-time traffic logging and analysis over IP networks. Its extensive database of over a thousand attack signatures is widely used nowadays. This signature-based approach entails one main drawback: only previously known attacks can be identified; mechanisms that can automatically "learn" patterns of 0-day attacks have yet to be developed.

In a similar way, knowledge-based MIDSs check network events against predefined rules or patterns of attacks. The goal is to employ representations of known attacks that are general enough to handle actual occurrences. The most common knowledge-based techniques applied to ID are expert systems, signature analysis, and state-transition analysis.

In machine learning methods for misuse detection [19], patterns and classes of attacks are automatically discovered rather than being pre-defined as in previously mentioned approaches, avoiding the costly handcrafting of profiles (via rules). It also avoids the labour-intensive process of carefully designing the applicable models needed for statistical approaches. The application of machine learning for anomaly detection was identified as a particularly promising research line [110]. Data mining can reveal strong associations between data elements, which often correspond to attack patterns. Classifiers then induce classes of attacks, based on data mining results and other attributes from training data with known attacks.

It has been pointed out [111] that data-mining-based ID usually relies on non-trivial assumptions regarding the availability and the quality of training data. One of the main problems is that, if the training dataset contains some intrusions, the IDS may assume that they are normal traffic and may fail to detect future instances of these attacks. This issue is addressed in [112, 113] and some authors [114, 115, 116, 117] propose clustering for unsupervised anomaly detection to overcome these problems.

Most statistical approaches to ID are anomaly-based and try to build profiles of "normal" user behaviour. To do so, they use a set of variables that may temporarily change to measure the "normal" behaviour of both systems and users. The goal is still to create representations of normal behaviour, as in the case of knowledge-based anomaly detection methods. The difference is that in this case the representations are generated automatically as parameters of pre-determined statistical models.

ANNs have been applied to the empirical development of network systems for traffic monitoring and ID, by using both supervised and unsupervised methods for automated detection. ANNs have been applied in a variety of ways to ID: on their own for the detection/classification of attacks, to pre-process input data for further processing, or as part of hybrid approaches.

Finally, one of the latest approaches to ID is based on immune systems. Although interesting and intuitively appealing [110], its application has proved difficult in practice [118].

3.2 Visualisation

The topic of visualisation for network security has been widely addressed since initial works in this field [44, 74, 297]. A great variety of visualisation-based approaches have been proposed. As this study focuses on visualisation of network data rather than network structure or topology, previous literature described in this section is limited to network data visualisation for NID. In these cases, the ID issue is enabled by providing a visual depiction of the network or the traffic. Thus, the identification of attacks must be performed through visual features as no alarms are triggered.

Different issues on user interfaces for ID are discussed in [119]. Additionally, the authors gave a general solution for the problems and challenges of such issues. [120] reports on activities to gather user requirements and to design an information visualisation tool for ID.

A broad compilation of visualisation approaches to ID is presented in [45]. It also discusses some interesting topics such as how to create and how to defend a security visualisation system. A review of different visualisation techniques applied to network visualisation can be found in [121]. Some visualisation-based ID works presented in these two studies are analysed in this section, together with others.

Previous work involving visualisation related to computer networks [43, 121, 122, 123, 124, 125, 126, 127, 128] has emphasized graphics that depict network topology, performance, connectivity, or bandwidth usage. Some other works propose the visualisation of alarms generated by IDSs to assist security personnel when discriminating false alarms:

- NIVA (Network Intrusion Visualisation Application) [129] integrates IDSs (Real Secure and Black Ice) with 3D visualisations to interactively investigate and detect structured attacks across both time and space.
- A practical guide of ID alarm visualisation is contained in [130]. This work describes basic chart-based visualisation techniques such as pie and bar charts that can be generated by means of the Visual Insights Advizor tool.
- SnortView [131] is a tool developed for analysing Snort logs and syslog data. It presents an updated 2D view every two minutes and shows four hours worth of alert data. A grid with time as the horizontal axis and source IP address as the vertical axis forms the main visualisation component.
- IDS RainStorm [132] visualises the alarms generated by StealthWatch in a matrix by using time and destination IP addresses. Colour is used to show the severity of alarms.
- [133] proposes two bi-dimensional matrices of IP addresses to work in unison with Snort. It is aimed at identifying and analysing automated attacks, such as the Welchia and Sasser worms. The colour of each pixel in the matrix is chosen regarding the number of the triggered Snort alerts. Snort alerts were split into 8 groups in the order of the alert number and a different colour was assigned to each group.

- [134] proposes the use of information visualisation to provide insight into the operation of some IDSs and to help the operator monitor the security state of a system.
- VisAlert [135] integrates logs and alerts from various sensors, IDSs, and hardware in an advance visualisation technique. The proposed visualisation concept tries to provide users with three different dimensions of alert information: when, where, and what.
- Hierarchical visualisation is applied in [47] to view IDS logs.
- ACID (Analysis Console for Intrusion Databases) [136] is intended to search and process a database of security events generated by various IDSs, firewalls, and network monitoring tools. Different charts are generated over these data. At present, it is devised for active analysis of Snort logs.

From a different standpoint, MOVICAB-IDS visualises traffic network data for the detection of intrusions.

3.2.1 Visualisation Techniques

Taking into account the applied visualisation techniques, previous work on NID can be categorised as specified in the following sections.

3.2.1.1 Scatter Plots

- 3-D scatter plots have been proposed in [137] to detect attacks such as Distributed Denial of Service (DDoS) and scanning activities. Each dot in the plot represents a flow, being defined as a set of packets sharing the same source IP address, destination IP address, and destination port number (i.e., packets in the same TCP connection). Attacks are identified as geometrical forms (parallel rectangles and lines) in such visualisation employing source IP addresses, destination IP addresses, and destination port numbers as axes. The attacks that are analysed occupy a visibly significant space in this 3D plot due to their inherent need for evasion manoeuvre (DoS attack) or wide scanning in a short amount of time (network/port scans), whereas "normal" (legitimate) flows only occupy a single point. One of the main drawbacks of such plots is the poor discrimination between attacks and "normal" traffic.
- The Spinning Cube of Potential Doom [138] plots TCP connections as coloured dots in a 3D spinning cube. The X and Z axes represent local and global IP addresses, while the Y axis depicts port numbers. Successful connections are shown as white dots while incomplete TCP connections are represented as multi-coloured dots with the colour varying by port number. Port scans appear as vertical lines, while the horizontal lines represent network scans. Combinations of them can also be visualised.
- The Shoki Packet Hustler [139] is an open source project that is able to plot any three user-specified variables (rather packet-extracted or derived) against each other simultaneously. Two different scatter plots are generated for such variables: three 2D plots (X-Y, X-Z, and Y-Z) and a 3D isometric view.

3.2.1.2 Parallel Coordinate Plots

- VisFlowConnect-IP [343, 348, 349] uses parallel coordinates to visualise incoming and outgoing network flows between internal and external hosts. The Parallel Axes View consists of three vertical parallel axes and the thickness of a line between two points represents the traffic volume between the hosts. Port information is also codified by means of colours. Additionally, data are filtered by a user-configurable time window.
- ID through parallel coordinate plots is widely covered in [45]. Visualising different features extracted from packet headers, sample attacks are identified: port scans generated by means of Nmap and Unicornscan, attacks through Nessus and Metasploit, and "real" attacks from a home ISP connection.
- PCAV (Parallel Coordinate Attack Visualisation) [44] makes use of parallel coordinate plots to detect unknown large-scale Internet attacks. Additionally, graphical signatures for such visualisation are defined in order to detect nine different attacks automatically.

3.2.1.3 Charts

- By analysing and reporting on data exported by routers, Flowscan [140] can reveal certain anomalies (e.g. DoS attack) due to an abrupt increase in the number of flows. Chart images are generated by RRGrapher to provide a continuous, near real-time view of the network traffic. Metrics such as flows-per-second, packets-per-second, and bytes-per-second over a default 48-hour period are represented as traditional X-Y line charts. The standard charts provide views of the traffic by network or subnet, by application or service, and by autonomous system.
- Stacked histograms of aggregate port activity are proposed in [141]. The horizontal axis maps time and the vertical axis maps interval values of packet count and total bytes. Colours are used to distinguish the different intervals in the vertical axis.
- 2D conditional histograms are built in [142] to detect distributed scans. This work is focused on parallelization strategies for building conditional histograms, offering significantly improved performance. Different histograms can be visualised: daily counts of unsuccessful incoming TCP connections directed at a certain port over a 365-day time period, connection counts at hourly resolution, or per-hour number of unsuccessful connection attempts over a two-day period.
- Simple horizontal bar charts (packet length vs. packet number) are used in [45]. As this technique does not by itself allow certain attacks to be clearly identified, it is enhanced by the *binary rainfall* technique [143].
- NetBytes Host Viewer [144] is an interactive visualisation tool designed to show the historical network flow data per network entity over time, using 2D and 3D charts. The entity could alternatively be a host (per port aggregation of data), a subnet (per host aggregation of data), or a larger network (per subnet aggregation of data). The list of entities (ports, hosts, or subnets) appears along

the Z axis of the display while time is displayed on the X axis. The third dimension (Y) represents the magnitude of traffic (in flows, packets, or bytes) seen by the entity in the time unit. Colour is used to specify all the data over time for a certain entity.

3.2.1.4 Graphs

- GrIDS (Graph-based IDS) [145] puts together reports of incidents and network traffic into graphs. It is able to aggregate those graphs into simpler forms at higher levels of the hierarchical decomposition of the protected organisation. It constructs graphs which represent hosts and activity in a network. Graphs, which can be of a variety of different types, are defined to identify different intrusions. For each type of graph, a set of rules are defined that specify how to build nodes, edges, attributes of nodes, attributes of edges, and global attributes of the graph.
- VisFlowCluster-IP [146] is a static off-line visualisation tool using graphs to highlight clusters of hosts closely related to each other. Each host in the network is modelled as a node in the graph. An edge is added between two nodes if there is certain relationship between them, i.e. they communicate with each other or they both access the same outside server(s). The arrangement of nodes within the graph is initially configured according to the host IP addresses and subsequently an algorithm is run that takes "attraction" and "repulsion" forces into account. Colour is assigned to nodes by considering the IP addresses. Finally, a clustering technique is applied to such graphs in order to identify groups of closely related nodes.
- A visualisation from a service-oriented perspective is proposed in [147]. In this graph-based visualisation each node in the network is represented as a compound glyph (node in the graph) giving details on the network node activity based upon its service usage. The size of the glyph represents the total amount of activity on the node, while the angled regions (in different colours) within the glyph show the percentage of the total activity related to a particular service. Activity links exist between nodes in the graph. Both appearance and location of the graph nodes depend on the type of the host they represent: hosts in the managed network or outside it. Temporal activity is visualised using time slicing techniques in the compound glyph.

3.2.1.5 Self-Organizing Maps

The connectionist visualisation approach to ID is not a new one; ANNs have been widely applied to the ID task but mainly from a classification-based approach using supervised learning. On the contrary, SOMs (see section 2.5.13) have not only been applied to classify network traffic but also to visualise it. This neural model is based on unsupervised learning and consequently does not rely on any prior knowledge of the data being analysed. Examples of studies in this field are:

- One of the initial attempts of using SOMs for ID [148] proposes visualising the user behaviour during a certain period in a simple way. The amount of traffic data is reduced by selecting a set of features that characterises the behaviour of the users in the network, forming a daily fingerprint of the network user. Although this amount of data is reduced in the feature selection process, it still is high-dimensional. To deal with these multidimensional data, the SOM is used to visualise the user behaviour in 2D. The maps are constructed using the whole dataset for the whole period: the data for all the users for all the days in the period. The chosen grid dimensions (18x14) were selected after empirically checking the accuracy of the mapping. The hexagonal (six-neighbour) lattice and the bubble neighbourhood function are used.

- [149] proposes a SOM-based visualisation focused on the information stored in event logs. The map is organised by similarity of events (considered as multi-dimensional vectors) rather than by network topology: events that are similar are close together in the display. New (or anomalous) user activities are identified by visual comparison. The neurons are portrayed as squares within a 2D grid and several visual features are depicted for each neuron according to certain values: foreground colour (value of the weight for the selected attribute), size (number of events mapped in the unit), and background colour (mean quantization error of all the entries resulting in the unit). The user interface provides the possibility of showing the map coloured by different attributes. In order to spot areas of similar nature across several attributes, the user can visually compare their distribution on the map to detect boundaries and correlated events.

  Previous work by this author [150] combines SOMs with the spring layout algorithm, a method with some scalability problems that was inspired by research in the field of physics.

- [151] applies the SOM, together with some other exploratory multivariate analysis methods and visualisation techniques. The SOM results are visualised by plotting a stars plot of the representative at each point in the hexagonal lattice, thereby combining the output of SOM and star plots together in a single diagram.
- A SOM is used in [152] to model and to visualise the relationship between known and unknown attacks in the KDD Cup 1999 dataset [153, 154]. A Gaussian neighbourhood function over a hexagonal 10x10 lattice is selected. After training and building the map, a hit score matrix is generated from the class labels (attack types) of the dataset, in which the neurons are labelled with the class label that has the highest hit score.
- [155] proposes a first step of traffic visualisation by applying a SOM. Once the SOM is trained, the weight information is used as an input to a Back Propagation Network for classification.

3.2.1.6 Advanced Interfaces and Combined Techniques

- SeeNet [43] includes three different graphical tools for network monitoring. Apart from link maps (that picture the connectivity of a network over a geographical map) and node maps (that display node-oriented data by showing a glyph over a geographical map), the other visualisation (the matrix display)

depicts network traffic on a coloured grid. This matrix-based visualisation concentrates on the links of a network, in an attempt to overcome some of the constraints associated with a geographical visualisation. The grid shows the volume of traffic between nodes in a network, each point representing the amount of traffic between a traffic source and a traffic destination by means of different colours. This visualisation relies on the order in which the nodes are assigned to rows and columns. The nodes can be arranged in approximate geographical order with west-to-east along the horizontal axis and correspondingly along the vertical axis.
- In [156], the previously developed SHADOW IDS is upgraded by different visualisation techniques, including pixel images (matrices), parallel coordinate plots, and colour histograms. Parallel coordinate plots use source and destination ports, source and destination IP addresses, and time as axes. A 256x256 colour histogram is built with the columns corresponding to the third octet of the local IP address and the rows to the fourth octet.
- Nam [157] is proposed as a network animator providing packet-level animation, protocol graphs, traditional time-event plots of protocol actions, and scenario editing capabilities. To provide an animated network traffic visualisation, Nam interprets a trace file containing time-indexed network events. This animation is based on a graph adjusted to the network topology. Once this graph is built, trace events indicate when packets enter and leave links and queues of the hosts in the network. Additionally, time-event graphs represent protocol-specific information.
- Several 3D displays using numerous visual attributes (shape, position, motion, size, orientation, colour, transparency, texture, and blinking) of geometric objects are proposed in [158], some of which are prototyped in the 3D Virtual Reality Modelling Language.
- Animated glyph-based visualisations are proposed in [107, 108] as an extension of previous work by the authors of [159]. These visualisations show connections from external hosts to a monitored server or small network environment. Through these visualisations, systems are represented as glyphs incorporating visual attributes representative of parameters in the data, such as number of users, system load, status, and unusual or unexpected activity. Initial connection requests are represented as parallel lines and hashes in such lines represent the number of user connections from one system to another (a single hash mark represents each user). Protocol information is codified in the style of the lines: solid, dashed, long dashed, solid with many arrows, etc. Colours are used to indicate the status of the activity: unusual or unexpected, questionable, etc.

  These visualisations are complemented in [160] with certain techniques that are more specifically geared towards the representation of behaviour: simple line-based histograms, pixel-based histograms, and a summary display showing the overview of all activity represented in the pixel-based histogram representation.

- The solution proposed in [161] combines visualisation and data mining of routing data to detect anomalies, intrusions, and router instability. ID is performed by combining star coordinate plots and a previously developed tool called

PaintingClass. Additionally, scatter plots of original data dimensions are also used.

- Visually fingerprinting the most common attack tools is proposed in [48]. To do so, parallel coordinate plots show traffic patterns between various hosts on a network. They extend the VisFlowConnect system [333, 338] by applying additional dimensionality, alternative encoding techniques, real-time packet capture, and focused attack tool specific data. Additionally, some other made-to-measure and scatter plots are proposed to complement the parallel coordinate plots.
- PortVis [162] is focused on TCP and UDP port traffic and produces visualisations of network traffic using 2D plots. The interface combines several components: a timeline of network activity, a primary display of network ports, a detailed view of selected ports, and charts of individual port activity. The main component employs a two-dimensional 256x256 matrix on which each position corresponds to a port; the vertical axis corresponds to the remainder of the port number divided by 256 and the horizontal axis corresponds to the integer value of the port number divided by 256. Network activity is summarised at each location in the plot using colour.
- NVisionIP [163] integrates different views of network data:
- Galaxy view: the overall state of a class-B network is shown in a scatter plot with IP addresses on each axis (subnet versus host). Each point represents the traffic destined for a certain host, and its colour shows the number of active ports at a given time.
- Small multiple view: bar charts are shown for a selection of specific ports of the machines on a given subnet.
- Machine view: all the details of a specific machine are presented by means of bar charts.

The way in which information from NVisionIP and VisFlowConnect can be correlated and used to identify several network anomalies is detailed in [164]. The NVisionIP system-oriented view is complemented with the VisFlowConnect network view in an intuitive way.

- Although proposed as a firewalling tool, the VisualFirewall [165] incorporates network monitoring and ID facilities. Two of its four different visualisations of network state focus on overall network security:
- The Visual Signature view shows packet flows as lines on a two-parallel axis plot. The right axis shows the global IP address space and the left axis shows ports on the local machine, using a cube root scale. A line from the local port to the foreign IP address shows that there is an exchange of packets between the local host and a foreign host. Its colour represents the transport protocol being used and its brightness depends on time: brighter lines correspond to newer packets and dull lines correspond to older packets.
- The Statistics view displays the throughput of the network (in bytes per second) on a line chart. The horizontal axis representing time is compared against three different measures of network traffic: overall, incoming, and outgoing throughput.

- [166] proposes several three-dimensional visualisations each one emphasizing different aspects of the data. Two visualisations based on Snort alarms are proposed in unison with a network data visualisation. The later being intended to support the identification of false positives generated by Snort. The visualisation of network traffic (referred as Island) is an iconic approach using a tree metaphor. Trees are built for each destination port number and branches represent the associated destination IP addresses. Similar to the branch idea, leaves are drawn at the top of the tree based on the components of the associated source IP addresses. The octets of the IP addresses affect the colour, the angle (about the tree), and the distance from the tree to the leaves. Finally a sphere depending on the number of bytes transferred is drawn at the top of each tree.
- A two-level visualisation methodology for the detection of network scans is proposed in [167]. The authors propose a detailed view technique for displaying individual scans. It consists of a 256x256 grid where the horizontal and vertical axes are the third and fourth bytes of the destination IP addresses. The colour of each dot represents various metrics based on statistical information regarding the arrival time at that IP address. Additionally, to compare pairs of scans, wavelet scalograms are proposed. On the other hand, a graph visualisation is proposed to provide a high level view of a large set of scans. Each node represents a scan, and the connection between any two nodes is weighted according to the wavelet comparison between them (0% means "completely different" and 100% means "total matching").
- [168] proposes a visualisation tool for real-time and forensic network data analysis. This tool combines link analysis using parallel coordinate plots with the time-sequence animation of scatter plots in a 2D/3D coordinated display. A 2-axes parallel coordinate plot is built by using source IP addresses and destination port numbers in which colours of lines denote the protocol (UDP or TCP). Glyphs based on temporal information are also built and move within the visualisation as the packets get older.
- Portall [169] correlates network traffic to the host process generating it through diagrams. Its main window shows a client-server layout of communications from the perspective of the monitored machines. Neutral, complementary colours and subtle shading gradients were used for the host and process markers, while contrasting primary colours were used for the port boxes and connection lines. Port activity bar charts are included in the process boxes.
- IDGraphs [170, 171] combines different visualisation techniques. It generates scatter plots of time versus unsuccessful connection, offering several aggregation methods that help reveal different attack patterns. These plots are also considered as time-series histographs featuring a time-versus-failed-connections mapping. Users can view a great number of such time series simultaneously supporting the discovery of attack patterns. Visual features such as pixel luminance are used to provide the user with further information. IDGraphs also includes further facilities such as an interactive query interface, a correlation analysis and in-depth examination supported by a zooming interface.
- The main visual component of TNV (Time-based Network traffic Visualizer) [172] is a matrix display showing network activity of hosts over time. Time is

shown in the horizontal axis and all available IP addresses are displayed on the vertical axis. The colour of each cell depends on the number of packets for that time interval. Connections between hosts are superimposed on the matrix as links, in a similar way to parallel coordinates; the colour of such links shows the associated protocol. This view is complemented by others such as: a histogram of the temporal network traffic activity, a diagram of port activity and the details of the raw packets.

- An interface consisting on different views is proposed in [173]. The timeline visualisation is a scatter plot of the entire time range available, combining time and port numbers (summarized in 32 ranges). This visualisation also includes a histogram showing the relative frequencies of each activity level (number of sessions) over the entire range of time. The grid visualisation consists of a 256x256 matrix (one dot for each of the 65,536 ports) depicting the port activity during a given time unit. This visualisation considers the port number as a two-byte number: the horizontal axis represents the high byte of the port number, and the vertical axis represents the low byte of the port number. There is a scatter plot in which each point represents a particular port, and the axes correspond to two different metrics. As in the case of the timeline visualisation, a histogram is included in the scatterplot visualisation. The volume visualisation generates a 3D plot by adding time to the grid visualisation. Finally, a port visualisation is offered, showing all the data available that concerns a particular port. Time-varying metrics associated with a port number are depicted in five different line graphs.
- VIAssist (Visual Assistant for Information Assurance Analysis) [174] incorporates a suite of integrated visualisations to examine the same traffic dataset from multiple perspectives. It is intended to reveal patterns that would not be discovered through a single perspective. These visualisations include:

  - A time histogram of packets count.
  - A matrix of source and destination IP addresses.
  - Extended parallel coordinate plots.
  - Trees in which the analysed host is the root of the display, destination port numbers constitute the second level of the hierarchy and destination IP addresses are the leaves.
  - Bar charts where the horizontal axis depicts the number of packets, and the vertical axis depicts the destination IP address.

  VIAssist has been lately integrated [175] with a network flow analysis system, SiLK (System for Internet-Level Knowledge). The SiLK network traffic analysis tools are used by US-CERT, JTF-GNO, and other organisations to collect, store, and analyse network traffic as flows.

- NFlowVis [176] incorporates five analysis views: general network overview, integrated ID view, flow visualisation of attacker connections, detailed host view, and NetFlow records view. The general overview shows aggregated traffic and port usage through line charts, grouped line-wise pixel arrangements, and port activity through coloured matrixes. The ID view links IDS alerts with the full NetFlow records and displays the textual data. Within the flow visualisation view, the monitored network is mapped to a treemap visualisation [177]. This visualisation hierarchically maps the monitored network infrastructure to prefixes of various granularities and enlarges high-load entities. The local IP prefixes or addresses are represented in the treemap and the external hosts (and hence, the intruders) are arranged at the borders. Internal and external IP addresses are linked using splines parameterized with prefix information. Detailed histograms are provided in the host view.

- Isis [178] provides users with two linked visual representations of temporal sequences of network flow traffic: the timeline (a collection of histograms) and the event plot (similar to a scatter plot). In both representations, IP addresses act as categorical values for the vertical axis with time on the horizontal axis. Timelines show an aggregated value over all events (such as the minimum, maximum, or average of the number of packets or bytes), while event plots reveal the patterns of individual events. The size, shape, and colour of the markers in the event plots can be changed based on certain flow attributes.

3.2.1.7 Others

- In one of the earliest works in this field [179], an Ethernet source/destination traffic matrix is proposed to identify network worms. Source and destination host numbers (octal) in a segment are respectively represented on the vertical and horizontal axis.
- IDIOT (Intrusion Detection In Our Time) [180] makes use of coloured Petri Nets [181] to represent attack signatures. Petri Nets [182] can be defined as a graphical language to design, specify, and verify systems, the main drawback of which is its high computational cost associated with comparing the stored attack signatures and the log-registered events.
- Spicules (spherical glyphs capable of representing up to 360 variables) are proposed in [183] to show nodes. Under this 3D geometric approach, the volume of a spicule is proportional to a computed value representing the perceived security fitness of the node. Each spicule has several tracking vectors protruding from its surface that represent different attributes of the node. For those attributes which have minimum and maximum values, tracking vectors travel from the horizontal plane towards the positive vertical axis on the surface of the spicule as the value of the feature approaches its maximum. Attributes that have no bound on growth are represented by fixed vectors around the equator of the spicule which grow outwards as their values increase. The vectors around the spicule for a system may be used as a geometric signature of the state of the system, represented as angles for tracking vectors and magnitudes for fixed vectors. The geometric signature for a specific attack could be also added to the spicule for a system under normal operation, resulting in a visual of what that system will look like under attack. The spicule for a computer on a network is plotted in a 3D space with its IP address on the X and Z axis, coupled with a mathematical quantity derived from the spicules vector angles on the Y axis.

- A 3D visualisation to detect network intrusions in telnet traffic is proposed in [184]. As this work combines visualisation and information retrieval techniques, telnet data are fed into the Telltale document analysis system, where each session is considered as a single document. Once a number of sessions have been entered, another session can be used as a query against them to get a similarity score. By using a session with a known attack as the query, the group of previously entered sessions is then scanned for other instances of the attack. This is done for multiple groups of sessions, and then the resulting similarity scores are shown in a glyph-based 3D visualisation generated by the Stereoscopic Field Analyzer visualisation system [185]. The scores of similarity of each session for three different attacks are arbitrarily assigned to the three axes in order to build the 3D visualisation.
- Tudumi [186] is a log visualisation system to assist in the auditing of log files to detect intrusive and anomalous behaviour. It focuses on three main types of user activity: remote access of the server, logging onto the server, and switching user accounts. These data are represented as a series of layered concentric disks in a 3D space. The bottom disk displays information about user switching and the remaining disks represent network access and log-in information. Each user is represented by a textured node.
- In [187], the interactive visualisation of Border Gateway Protocol (BGP) data is proposed to identify anomalous situations, to characterise routing behaviour, and to identify weaknesses in connectivity. To do so, each IP address is mapped to one pixel on a 512x512 pixel square. A square is repeatedly subdivided into 4 equal squares: in mapping a 32-bit address to a square, the first two most significant bits of the address are selected to place it in one of the 4 sub-squares. The next two most significant bits are repeatedly used to place the prefix in a subsquare within a sub-square until the prefix is in a square with the size of a single pixel or the bits of the prefix are exhausted. A pixel is coloured yellow if a change in the routing data occurred on the current day and brown if a change occurred on a previous day. In the detail windows, a coloured rectangle is shown for each routing data change: its position is determined by the IP prefix, the size by the mask, the brightness by how long ago the change occurred, and the hue by the type of the change.
- The Immersive Network Monitoring system [188] employs 3D real world metaphors to lay out and encode abstract information to fully engage the human perceptual and cognitive systems. Using the concept of "territory", the internal address space of the monitored network is mapped into a circular region. The external address space is then mapped into a hemispherical region centred on the first region and separated by a shield (represented by a semi-transparent dome). The connections between internal and external hosts are presented by lines with particular colour and length, representing information such as service type, duration of that connection, and source/destination IP addresses.
- VISUAL [189] can be used for analysis of packet data for a subnet consisting of less than 1,000 hosts. A home-centric ("*us versus them*") perspective of the network is adapted: all home (internal) IP addresses are shown as a square grid with all these addresses represented as small squares. External IP addresses are

located along the screen: the first two bytes of an external IP address determine its X coordinate on the screen and the last two bytes determine the Y coordinate. As a consequence, external IP addresses from the same class B or C network are shown as vertical lines. Lines from individual computers in the home network communicating to other external hosts are also shown, which simply mean that there was some communication between the computers. Line colour shows the direction of traffic. With this home-centric visualisations, a user can see which home IP addresses received a great deal of connections (fan-in) and which external IP addresses received the most hits from the internal IP addresses (fan-out).

- The visualisation environment proposed in [190] employs a visualisation technique consisting of several concentric circles in a square. The local IP address is represented around the radius of the internal circles. Remote IP addresses are represented along the top and bottom of the square edges. The top edge is used if the local IP appears in the top semi-circle and the bottom edge is used if the local IP appears in the bottom semicircle. Similarly, (the right edge for the right semi-circle and the left edge for the left semi-circle) port numbers are represented on the left and right edges of the square. The circles are representative of the age of the identified activity: the outer circles represent the most current data while each inner circle represents data m time units older, where m is the circle number.
- IDtk [191] generates glyph-based visualisations in two or three dimensions. The user could choose the mappings between data attributes and visualisation features. In other words, users select the data features for the axes, and those determining the colour and size for each data item.
- NetViewer [192] provides the view of network traffic as a sequence of visual images. The normalized packet count of the 256 entries of each IP address byte is arranged in a 16x16 matrix for visual representation at the sampling point. The 16x16 matrix of each of the 4 bytes of IP addresses are organised as a frame for both source and destination IP addresses, respectively. Image processing techniques are proposed for the detection of attacks in such matrices.
- Network traffic is visualised in [193] by mapping its given one-dimensional time series into a co-occurrence matrix and applying statistical texture analysis methods from the domain of digital image processing. Some details depending on what is of specific interest (i.e. number of TCP streams or open TCP connections, or number of IP, TCP, UDP, or ICMP packets per time interval) can be used by selecting different aggregation functions such as the number of packets captured or a certain protocol.
- Through SVision [194], a network is represented as a community of hosts roaming in a 3D space which is defined by the set of services the hosts are using. It is intended to graphically cluster the hosts into normal and abnormal ones. Spheres representing the positions of hosts and colours (gray and black) are used to distinguish between the internal and external hosts, respectively. Additionally, the intensity of the colours changes from dark to light with respect to the time. A 2D plane is applied to discriminate between sparsely-active and constantly active hosts with respect to a selected set of services. For the

sake of clarity, SVision only highlights the hosts that might represent a potential threat to the network as it has problems in effectively dealing with large-scale attacks on a high-speed network. The "normal" hosts overlap near the centre of the view.

- Properties of packet flows are mapped to pixel coordinates using Space-filling Curves (SFCs) in [195]. SFCs generate traffic images that provide both storage and bandwidth savings in characterising traffic flows and identifying anomalous behaviour. Packets are characterised by four features: source/destination IP address and source/destination port. Packet records are aggregated to form a histogram representing the frequency of occurrence of each one of the possible four-tuple values. This histogram is converted to a one dimensional representation by means of SFCs. Then, a 2D representation for image visualisation is generated from the one dimensional representation.

3.2.2 Visualised Data

Taking into account the visualised data, previous work can by classified as specified in the following sections.

3.2.2.1 Packet Headers

- Raw packet data are summarized by GrIDS [145] into events consisting of a single network packet or a collection of packets. Connections are then defined as sequences of events labelled with source and destination IP addresses, source and destination ports, sequence number, and time.
- [149] visualises events based on the following data: timestamp, source and destination IP addresses, source and destination ports, packet length (UDP only), data sequence number of the packet (TCP only), data sequence number of the data expected in return (TCP only), number of bytes of receive buffer space, available indication of whether or not the data is urgent (TCP only), and flags like SYN, FIN, PUSH, and RST (TCP only).
- The extension of the SHADOW IDS propose in [156] mainly employs source and destination ports, source and destination IP addresses, and time.
- Individual packets are grouped into bidirectional network flows analogous to TCP connections in the Immersive Network Monitoring system [188]. A flow is identified by a 5-tuple: source/destination IP address, source/destination port, and protocol. For each flow, the following features are computed: start and end times, endpoint IP addresses, endpoint TCP or UDP ports (when applicable), and the total number of bytes and packets in each direction.
- The visualisation strategy proposed in [137] uses destination port numbers, source IP addresses, and destination IP addresses extracted from traffic flows.
- The Spinning Cube of Potential Doom [138] uses TCP connections (including every completed and attempted TCP connection) collected through the Bro IDS [196, 197]. These connections are characterized by source IP address, destination IP address, and destination port number.

- The following data are visualised in [48]: source and destination IP addresses, source and destination ports, and protocol type (TCP or UDP, inbound or outbound from home networks). Additionally, time is also used in some of the plots proposed in this work.
- Heavily sanitised data are analysed in PortVis [162]. Network data are reduced to a set of counts of entities such as amount of TCP sessions or amount of different source IP addresses that are present. From this general information aggregated from network traffic flows, the following pieces of data, accumulated over each hour, are used: protocol (TCP or UDP), port, hour, session count, unique source/destination IP addresses, unique source/destination IP address pairs, and unique source countries.
- VISUAL [189] uses a data source of packet traces, specifically pre-processed from pcap [198] files. IP addresses are mainly used to build the visualisation.
- By visualising network packet header data over time, the system proposed in [141] is able to see the penetration and subsequent activity of the Sasser worm. This data include port numbers, size, timestamp, and IP addresses.
- Two different data sources are proposed in [166]. The first one consists of alerts generated through Snort, while the second one is the 1999 DARPA dataset [199, 200]. The raw traffic data are aggregated according to destination port numbers and source/destination IP addresses. For each aggregated element, packet size and number of packets are considered.
- The tool in [168] uses source IP addresses, destination port numbers, protocol information, and time of each packet.
- NetViewer [192] passively monitors packet headers of network traffic at regular intervals. The gathered data include source/destination IP addresses, source/destination port numbers, traffic volume in bytes, packets, and other information.
- TNV [172] proposes an advanced interface considering several pieces of information: host IP address, packets timestamp, traffic volume at each host/port, and connections between hosts. Additionally, there is a view showing header information of the captured packets.
- Sanitised data are visualised in [173] to identify network and port scans. This data is reduced to a set of entity counts such as TCP sessions or IP addresses. It includes protocol label, port number, and time information for filtering and positioning the data. Additionally, attribute values are also used: session count, unique source addresses, unique destination addresses, unique source-destination address pairs, and unique source countries.
- SVision [194] was tested on DARPA datasets. Six fields from the packet headers were used: source and destination IP addresses, source and destination port numbers, packet length, and protocol type.
- In [195], the following fields of packet headers are used: source IP address, destination IP address, source port, and destination port.
- The system proposed in [147] uses source/destination IP address and source/destination port numbers from network packets captured at each machine on the network.

- PCAV [44] displays network flows by using the source and destination IP addresses, the destination port, and the average packet length in the flow. A flow is a single network connection and can consist of millions of packets.

3.2.2.2 NetFlow Data

- Flowscan [140] analyses NetFlow data, providing a visualisation in the units of five-minute intervals. It can plot packet, byte, and flow count by IP protocol (such as ICMP, TCP, and UDP), service or application, IP prefixes (class A, B, or C network), or autonomous system pairs.
- NVisionIP [163] visualises NetFlow data. The Galaxy View is based on IP addresses while the Small Multiple View uses port numbers to visualise the data.
- IDGraphs [278, 279] allows the visualisation of NetFlow connection records taking into account the number of unsuccessful connections. Its main visualisation facility aggregates streams of packets by source IP address and destination port number, destination IP address and destination port number, source IP address and destination IP address, destination IP address, source IP address, or destination port number. According to the authors, the number of failed connections (SYN-SYN/ACK) is a strong indicator of suspicious network flows.
- VisFlowCluster-IP [146] uses NetFlow data from two different sources. For each record, the following information is analysed: source and destination IP addresses, source and destination ports, number of bytes and packets, start and end times, and protocol type.
- NetBytes [144] uses application volume data obtained from multisets produced by the SiLK analysis tools based on NetFlow data. SiLK tools are applied to calculate counts about flows, bytes or packets for specific unique key identifiers such as source ports, destination ports, source IP addresses, and destination IP addresses.
- NFlowVis [176] stores NetFlow data (original features and some other derived such as flow count, transferred packets or bytes, etc.) in a relational database system and links these flows to alerts from IDSs or public warnings.

3.2.2.3 Others

- The matrix display of SeeNet [43] employs the list of nodes in a network and different statistics on the associated links. Several statistics such as absolute/percentage overload can be used.
- The user account logs of more than 600 users for a period of 400 days are used in [148]. These logs give information on the processes performed by the users, including CPU-times, characters transmitted, and blocks read. A feature reduction process is performed over these logs, reducing the 34 initial features to only 16.
- The trace files used by Nam [157] contain information on static network layout and on dynamic events such as packet arrivals, departures, drops, and link failures. Nam's animation component displays only a subset of the data in the trace

files. Other Nam components handle additional information such as packet headers or protocol state variables.

- In [184], the tcpdump [198] data of the 1998 DARPA dataset [1] are used. As only anomalous telnet traffic is targeted, telnet packets (packets that involved port 23 as either the source or destination port) were selected. For each one of these packets, the following features were considered: timestamp, protocol, source IP address, source port, destination IP address, and destination port. This information was subsequently aggregated as telnet sessions.
- The visualisation tool proposed in [69, 70] uses data from the Hummer IDS in order to represent network events between a monitored system and other hosts. This includes normal log files with additional statistics and other available system information. Data such as amount of users, status of the activity, protocol, and source/destination host are used to build the visualisation up.
- Tudumi [186] is based on log files and does not filter out any log information. Instead, it represents all kinds of events in the log at anytime. Access host information is summarized on the basis of its domain name.
- In [187], data from the Border Gateway Protocol are visualised by a diagram based on IP addresses.
- The 41 variables in the KDD Cup 1999 dataset, comprising both the training set (494,021 connections) and the test set (311,029 connections) were used in [161].
- VisFlowConnect-IP [110, 111, 112] employs the following features extracted from NetFlow records: source and destination IP addresses, protocol information, timestamp, and the amount of packets within a flow.
- Portall [169] uses information on hosts and processes. The Windows host monitor gathers process and port information via the IP Helper Application Programming Interface (IPHLPAPI). This interface retrieves all open TCP and UDP ports via a pair of undocumented Windows system calls. Additionally, the host monitor writes timestamped entries containing process and port information in the database. On the other hand, the Windows network monitor collects (by means of the WinPCap packet capture library) and stores packet header data and timestamps in the database.
- Incorporating firewalling, network monitoring, and ID facilities, the Visual-Firewall [165] uses both firewall and IDS alarm data. It also utilizes network data to provide simultaneous representations of relevant information. This network data include: foreign IP addresses, local port numbers, protocol, time, and packet counts (incoming/outgoing/overall).
- One of the 3D visualisations proposed in [166] is applied to the 1999 DARPA dataset. The other two employ data about the alarm data generated by Snort.
- Both packet- and connection-based data can be used in the visualisation environment proposed in [190]. Regardless of the data format, the visualisation is based on IP addresses, port numbers, and time.
- IDtk [191] can visualise raw TCP packet data or alerts generated by IDSs such as Snort. Due to its versatility, different features of TCP packet data such as source and destination IP addresses or destination port and time can be chosen.

- Network connection data collected by Bro at NERSC (National Energy Research Scientific Computing Center, Berkeley Lab) are used in [142]. This dataset contains 2.5 billion records, each with 22 attributes such as source and destination IP addresses, source and destination ports, start time, duration, and number of bytes sent along with additional network connection information.
- The KDD Cup 1999 dataset is used in [152]. To overcome some of the problems of this dataset, it was resampled: 1,000 samples from each class were randomly selected. Those classes having less than 1,000 samples were over sampled (with duplicates).
- [155] also uses the KDD Cup 1999 dataset. The dataset is pre-processed to remove redundancy, non-numerical attributes are represented in numerical form, and data are normalised.
- VIAssist [174] supports multiple data sources such as IDS alert databases, network flow data, or incident repositories. Flows are aggregated according to TCP sequence number, or a time threshold for UDP communications between unique host pairs. The different visualisations within VIAssist use source and destination IP addresses, source and destination ports, and packet counts.
- The data used by Isis [178] are the routed ICMP, UDP, and TCP network flows, captured by a sensor at the network gateway. Each flow (defined by the protocol, source/destination IP addresses, and source/destination ports) summarizes the time, duration, packet counts, and byte counts of a network connection at the transport layer.
- In the Shoki Packet Hustler [139] all the variables in the standard IP, TCP, UDP and ICMP headers are available for visualisation. Additionally, some other statistics such as relative times or running average can be visualised.

3.3 Agents and Multiagent Systems

As previously mentioned, IDSs have evolved from monolithic batch-oriented systems to distributed real-time networks of components. As a result, new paradigms have been designed to support such tools. Agents and MASs (described in section 2.6) are one of the paradigms that best fit this setting. Furthermore, some other AI techniques can be combined with this paradigm to make more intelligent agents.

Previous attempts to take advantage of agents and MASs in the field of ID include:

- JAM (Java Agents for Metalerning) [201] combines intelligent agents and data mining techniques. When applied to the ID problem, an association-rules algorithm determines the relationships between the different fields in audit trails, while a meta-learning classifier learns the signatures of attacks. Features of these two data-mining techniques are extracted and used to compute models of intrusion behaviour.
- In the 90s, DARPA defined the Common Intrusion Detection Framework (CIDF) as a general framework for IDS development. The Open Infrastructure [202] comprises a general infrastructure for agent based ID that is CIDF compliant. This infrastructure defines a layered agent hierarchy, consisting of the

following agent types: Decision-Response agents (responsible for responding to intrusions), Reconnaissance agents (gather information), Analysis agents (analyse the gathered information), Directory/Key Management agents, and Storage agents. The two last-named agents provide support functions to the other agents.

- AAFID (Autonomous Agents For Intrusion Detection) [203] is a distributed IDS architecture employing autonomous agents, which are defined as "*software agents that perform a certain security monitoring function at a host*". This architecture defines the following main components:
  - Agents: monitor certain aspects of hosts and report them to the appropriate transceiver.
  - Filters: perform data selection and function as an abstraction layer for agents.
  - Transceivers: act as external communications interfaces for hosts.
  - Monitors: the highest-level entities that control entities that are running in several different hosts.
  - User interfaces: interact with a monitor to request information and to provide instructions.

  This architecture does not detail the inner structure or the mechanisms of the proposed agents that use filters to obtain data in a system-independent manner. In other words, these agents do not depend on the operating system of the hosts. Additionally, AAFID agents do not have the authority to directly generate an alarm and do not communicate directly with each other.

- In [204], a general MAS framework for ID is also proposed. Authors suggest the development of four main modules, namely the sniffing module (to be implemented as a simple reflex agent), the analysis module (to be implemented as several agents that keeps track of the environment by examining past packets), the decision module (to be implemented as goal-based agents to make the appropriate decisions), and the reporting module (to be implemented as two simple reflex agents: logging and alert generator agents).

  The sniffing agent sends the previously stored data to the analysis agents when the latter request new data. One analyser agent is created for each one of the attacks to be identified. They analyse the traffic reported from the sniffing module, searching for signatures of attacks and consequently building a list of suspicious packets.

  Decision agents are attack dependant. They calculate the severity of the attack they are in charge of monitoring from the list of suspicious packets built by analyser agents. Decision agents also take the necessary action according to the level of severity.

  Finally, the logging agent keeps track of the logging file, accounting for the list of suspect packets generated by the decision agents. Subsequently, the alert generator agent sends alerts to the system administrator according to the list of decisions.

- Some topics regarding ID based on MASs are briefly discussed in [205], where a general framework for ID is proposed. Such framework includes the following classes of agents: learning data management agents, classifier testing agents, meta-data forming agents, and learning agents.
- SPIDeR-MAN (Synergistic and Perceptual Intrusion Detection with Reinforcement in a Multi-Agent Neural Network) is proposed in [206]. Each agent uses a SOM and ordinary rule-based classifiers to detect intrusive activities. A blackboard mechanism is used for the aggregation of results generated from such agents (i.e. a group decision). Reinforcement learning is carried out with the reinforcement signal that is generated within the blackboard and distributed over all the agents which are involved in the group decision making.
- A heterogeneous alert correlation approach to ID by means of a MAS is proposed in [207]. In this study alert correlation refers to the management of alerts generated by a set of classifiers, each of them trained for detecting attacks of a particular class (DoS, Probe, U2R, etc.). Although it is a HIDS, the main idea underlying the design of this MAS could also be applied to NIDSs. In accordance with Gaia methodology (described in Chapter 4), roles and protocols are specified in this study. The roles are mapped into the following agent classes:

  - NetLevelAgent (DataSensor role): in charge of raw data pre-processing and extracting both events and secondary features.
  - BaseClassifiers (DecisionProvider role): performs source based classification and produces decisions after receiving events from sources. Several subclasses are defined to cover the different predefined types of attacks and the different data sources.
  - Metaclassifiers (DecisionReceiver and DecisionProvider roles): one agent of this class is instantiated for each one of the attack types. They combine decisions produced by the BaseClassifiers agents of the assigned attack type.
  - SystemMonitor (ObjectMonitor role): visualises the information about security status.

- CIDS (Cougaar-based IDS) [208] provides a hierarchical security agent framework, where security nodes are defined as consisting of four different agents (manager agent, monitor agent, decision agent, and action agent) developed over the Cougaar framework [209]. It uses intelligent decision support modules to detect some anomalies and intrusions from user to packet level. The output of CIDS (generated by the Action Agent) consists of the environment status report (IDMEF format [210]) as well as recommendations of actions to be taken against the ongoing intrusive activities. The system employs a knowledge base of known attacks and a fuzzy inference engine to classify network activities as either legitimate or malicious.

- PAID (Probabilistic Agent-Based IDS) [211] is a cooperative agent architecture where autonomous agents perform specific ID tasks (e.g., identifying IP-spoofing attacks). It uses three types of agents:

  - System monitoring agents: responsible for collecting, transforming, and distributing intrusion specific data upon request and evoking information collecting procedures
  - Intrusion-monitoring agents: encapsulate a Bayesian Network and performs belief update using both facts (observed values) and beliefs (derived values). They generate probability distributions (beliefs) over intrusion variables that may be shared with other agents, which constitutes the main novelty of PAID. Methods for modelling errors and resolving conflicts among beliefs are also defined.
  - Registry agents: coordinate system-monitoring and intrusion-monitoring agents.
  - A multiagent IDS framework for decentralised intrusion prevention and detection is proposed in [212]. The MAS structure is tree-hierarchical and consists of the following agents:
  - Monitor agents: capture traffic, pre-process it (reducing irrelevant and noisy data), and extract the latent independent features by applying feature selection methods.
  - Decision agents: perform unsupervised anomaly learning and classification. To do so, an ant colony clustering model is deployed in these agents. When attacks are detected, they send simple notification messages to corresponding action and coordination agents.
  - Action agents: perform passive or reactive responses to different attacks.
  - Coordination agents: aggregate and analyse high-level detection results to enhance the predictability and efficiency.
  - User Interface agents: interact with the users and interpret the intrusion information and alarms.
  - Registration agents: allocate and look up all the other agents.

- A MAS comprising intelligent agents is proposed in [213] for detecting probes (scans). These intelligent agents were encapsulated with different AI paradigms: support vector machines, multi-variate adaptive regression, and linear genetic programming. Thanks to this agent-based approach, specific agents can be designed and implemented in a distributed fashion bearing in mind prior knowledge of the device and user profiles of the network.

  By adding new agents, this system can be adapted to large networks. Due to the interaction of different agents, failure of one agent may not degrade the overall detection performance of the network.

  In addition to the above works, others have focused on the mobile-agents approach: agents which travel along the different hosts in the network that is monitored. Some issues concerning the application of mobile agents to ID are further discussed in [214], and examples of this approach are:

- IDA (ID Agent system) [215] is aimed at detecting many intrusions efficiently rather than accurately detecting all intrusions. To do so, it approaches ID from a novel standpoint: instead of continuously monitoring the activity of users, it watches events that may relate to intrusions (MLSI - Mark Left by Suspected Intruders). When an MLSI is detected, IDA collects further information, analyses it and decides whether an intrusion has taken place. To do so, two kinds of mobile agents contribute to the information collection stage: a tracing agent is sent to the host where suspicious activity comes from and once there, it activates an information-gathering agent. Several information-gathering agents may be activated by several different tracing agents on the same target system.
- Micael [216] was proposed as an IDS architecture based on mobile agents. Its main difference with regard to previous proposals is task division. ID tasks are distributed to the following kinds of agents: head quarter (centralizes the system's control functions), sentinels (collect relevant information and inform the head quarter agents about eventual anomalies), detachments (implement the counter-measures of the IDS), auditors (check the integrity of the active agents), and special agents with different duties. By moving throughout the network, the mobile auditor agents can audit each of the defended hosts sequentially.
- Mobile agents are applied in [217] to make critical IDS components resistant to flooding DoS and penetration attacks. To do so, the attacked IDS component will be automatically relocated to a different (still operational) host. This relocation is invisible to the attacker who is then unable to continue the attack.

  Every critical agent has one or more backup agents (maintaining full or partial state information of the agent they are backing up) that reside on distinct hosts within the same domain. When the machine hosting a critical agent is down (for whatever the reason), its backup agents contact each other to decide on a successor that will resume the functions of the original agent.

  One of the main drawbacks of this solution is that the network may temporarily be partially unprotected while the IDS critical components are moving from one host to another.
- SPARTA (Security Policy Adaptation Reinforced Through Agents) is proposed in [218] as an architecture to collect and relate distributed ID data using mobile agents. According to the authors, SPARTA mobile agents enable distributed analysis, improve scalability, and increase fault tolerance. Some security issues about these mobile agents are considered.

  The required information (interesting host events) for event correlation is collected locally and stored, in what is considered to be a distributed database with horizontal fragmentation. Mobile agents are in charge of collecting the distributed information (matching a given pattern) to answer user queries.
- SANTA (Security Agents for Network Traffic Analysis) [219] is proposed as a distributed architecture for network security using packet, process, system, and user information. It attempts to emulate mechanisms of the natural immune system using IBM's Aglets agents. The proposed monitoring agents roam around

the machines (hosts or routers) and monitor the situation in the network (i.e., looking for changes such as malfunctions, faults, abnormalities, misuse, deviations, intrusions, etc.). These immunity-based agents can mutually recognize each other's activities and implement Adaptive Resonance Theory neural networks and a fuzzy controller for ID. According to the underlying security policies and the information from the monitoring agents, decision/action agents make decisions as to whether an action should be taken by killer agents. These killer agents terminate processes that are responsible for intrusive behaviour on the network.

- A distributed ID architecture, completed with a data warehouse and mobile and stationary agents is proposed in [220]. The MAS is combined with a rule generation algorithm, genetic algorithms, and datawarehouse techniques to facilitate building, monitoring, and analysing global, spatio-temporal views of intrusions on large distributed systems. System calls executed by privileged processes are classified after being represented as feature vectors. To do so, different agents are defined:

  - Data cleaner agents: these stationary agents process data obtained from log files, network protocol monitors, and system activity monitors into homogeneous formats.
  - Low-level agents: these mobile agents form the first level of ID. They travel to each of their associated data cleaner agents, gather recent information, and classify the data to determine whether suspicious activity is occurring. These agents collaborate to set their suspicion level to determine cooperatively whether a suspicious action is more interesting in the presence of other suspicious activity.
  - High-level agents: maintain the data warehouse by combining knowledge and data from the low-level agents. The high-level agents apply data mining algorithms to discover associations and patterns.
  - Interface agent: directs the operation of the agents in the system, maintains the status reported by the mobile agents, and provides access to the data warehouse features.

- In [221] a multiagent IDS (MAIDS) architecture containing mobile agents is proposed. These lightweight agents, located in the middle of the architecture, form the first line of ID. They periodically travel between monitored systems, obtain the gleaned information, and classify the data to determine whether singular intrusions have occurred.
- In the MA-IDS architecture [222], mobile agents are employed to process information in a coordinated manner from each monitored host. Only the critical components in the MA-IDS architecture (Assistant and Response agents) are designed as mobile agents. An Assistant mobile agent is dispatched by the Manager to patrol (gather information) the network. Assistant mobile agents are intended to determine whether some suspicious activities in different hosts are part of a distributed intrusion. If that is the case, the Manager component will possibly dispatch a Response mobile agent to respond "intelligently" to each monitored host.

It is claimed that these mobile agents are capable of evading attackers and even recovering from an attack. Additionally, agent mobility makes distributed ID possible by means of data correlation and cooperative detection.

- An interesting and comprehensive discussion about optimising the analysis of NIDSs through mobile agents is presented in [223]. The main proposal is to place the mobile analyser components of the NIDS closer together in the network and, if possible, to shift the processing load to underused nodes.
- APHIDS (Agent-Based Programmable Hybrid Intrusion Detection System) [224] implements the distributed search and analysis tasks by using mobile agents equipped with scripting capability to automate evidence gathering. This architecture is similar to the SPARTA and MAIDS systems (all of which exploit the mobility of the agents to perform distributed correlation), but APHIDS allows the specification of coordinated analysis tasks using a high-level specification language. Mobile agents are used for monitoring the output from other previously running IDSs (HIDSs or NIDSs), querying the log files and system state, and reporting results.
- APHIDS was subsequently upgraded, generating APHIDS++ [225] that introduces a two-level caching scheme:

  - Task Agents enter the first level cache mode (busy wait at the attacked machine) after having handled an initial attack. Each Task Agent maintains a publicly accessible queue of pending attacks to handle.
  - If no new attacks are sent to a Task Agent within a certain time limit, the agent enters the second cache level mode, in which it is flushed to its host machine's disk. Thus, resource consumption in the host machine is reduced.

  Some other improvements of APHIDS++ are the addition of an optional intelligent agent and an XML implementation of the Distributed Correlation Script.

- Two different agent classes are proposed in [226]: monitoring agents (AM) and managing agents (AZ). AM observe the nodes, process captured information, and draw conclusions for the evaluation of the current state of system security. AM agents can travel along the network to monitor different areas that may be at risk of attacks. On the other hand, AZ agents are responsible for creating profiles of attacks, managing AM agents, and updating its database and ontology.
- IDReAM (Intrusion Detection and Response executed with Agent Mobility) is proposed in [227] as a new approach to build a completely distributed and decentralized ID and Response System in computer networks. Conceptually, IDReAM combines mobile agents with self-organizing paradigms inspired by natural life systems: the immune system that protects the human body against pathogens and the stigmergic paradigm of a colony of ants. Those two natural systems display a "social life" in the organisation of their respective entities (immune cells and ants), which the author states is not possible without mobility. IDReAM is assessed in terms of resource consumption and intrusion response efficiency.
- IDSUDA (Intrusion Detection System Using Distributed Agents) [228] proposes the application of mobile agents to monitor the usage of various system

resources in order to detect deviations from normal usage. The behaviour of the attacker is tracked by following up the intruder's movements from one resource to another.

- A general distributed IDS framework based on mobile agents is proposed in [229]. Some of the components in this model are designed as mobile agents for the purpose of high adaptability and security of the system. It is claimed that these mobile agents can evade intrusion and recover by themselves when attacked, although no further explanations are provided on that point.

3.4 Novelties of the Proposed IDS

This section justifies the development of a novel IDS through a discussion of the different issues that relate to its scientific contribution. As stated in section 2.1, it is clear that IDSs have become a required element in addition to the computer security infrastructure of most organisations. A few years ago, NIDSs started to become a standard part of network infrastructure, just as firewalls had done in previous years [20]. Although prevention is becoming more and more effective, it will never eliminate the need for IDSs and network monitoring [20].

Great effort has been devoted to the ID field, but several issues concerning IDS design, development, and performance are still open for further research. Most MIDSs rely on models of known attacks. Thus, the effectiveness of these systems depends on the "goodness" of their models; if a model of an attack does not cover all the possible modifications, the performance of the IDS will be greatly impaired. Previous works have reported this issue [11, 142, 143, 144, 145, 146, 147]. Indeed, one of the main challenges facing ID research is this problem of identifying previously unseen (0-day) attacks.

On the other hand, NIDSs do still produce a large number of false alarms (specially false positives) [20] and generate a huge amount of record alerts and notifications as text logs. Analysing these logs and identifying false alarms are extremely monotonous and time-consuming tasks that are often performed manually. As a consequence, IDSs are not frequently monitored, what means they are virtually useless, even if correctly configured [20].

As stated in [16], scant attention has been given to visualisation in the ID field, although visual presentations do, in general, help operators and security managers, in particular, to interpret large quantities of data. Most IDSs do not provide any way of viewing information other than through lists, aggregates, or trends of raw data. They can generate different alarms when an anomalous situation is detected, broaden monitoring tasks, and increase situational awareness. However, they can neither provide a general overview of what is happening in the network nor support a detailed packet-level inspection [172].

Visualisation techniques take advantage of the outstanding capabilities of the human visual system to detect patterns and anomalies in visual representations of abstract data [230]. Another advantage of visualisation is that it transforms the task of analysing network data from a perceptually serial process (by reading textual data) to a perceptually parallel process by presenting more concepts [231]. According to this, the visualisation approach to ID implies several advantages:

- Attack visualisation can provide fresh insight into the analysed data, allowing the deduction of new hypotheses usually lost in complex analysis [44]. As a consequence, 0-day attacks can be easily detected.
- "*Visualisation tools need to be designed so that anomalies can be easily flagged for later analysis by more experienced analysts*" [11]. Visualisation for ID can help in training security personnel with no previous experience in security tasks, as well as reducing the time spent by more experienced personnel.
- For effective analysis, network data must be correlated with several variables. This requires dealing with highly heterogeneous, complex, and noisy data. The visualisation approach simplifies this problem by presenting the traffic situation in an intuitive way as visual images can give perceptual clues to the administrators [44].
- Attack visualisation can be much faster than other anomaly detection approaches [44]. As it also reduces the time and effort of reviewing security logs, it implies a great reduction of the time (and hence resources) required for ID.

Although many ID tools have begun to incorporate advanced graphical user interfaces, most of them fail to provide an intuitive and comprehensive visualisation of network traffic. To identify intrusions, one has to look for "interesting" structure and for "abnormal" or unusual data. Although this process can not be precisely detailed in a general and objective way, "one usually can recognize unusual data when one sees them" [23].

This work follows a line of research which, guided by previous work, builds on the following strong points:

- **Continuous inspection of network traffic by visualizing individual packets and not summarized/aggregated information**

  The underlying idea in the present study is not only to detect anomalous situations but also to visualise protocol interactions and traffic volume. As pointed out in [35], packet-based ID has several advantages:

  - There are numerous network-specific attacks (e.g. large distributed denial-of-service attacks) that cannot be detected by using audit information on the host but only by using information on network infrastructure.
  - TCP/IP standardization of network traffic facilitates collecting, formatting and analysing information from heterogeneous audit trail formats that come from different portions of large and complex networks.
  - By visualising connection-based data, important information, such as port scans and more importantly low and slow port scans can be missed [190, 232].

However, using network packets for ID also entails several drawbacks:

- When an intrusion has been detected, it is not straightforward to identify the attacker, since there is no direct association between network packets and the identity of the user who actually performed the attack.
- If the packets are encrypted, it is practically impossible to analyse the payload of the packets, as important information may be hidden from network sniffers.

- **Combination of different AI paradigms**

  Hybrid Artificial Intelligence Systems (HAISs) [232, 233, 234, 235] combine different AI paradigms in order to build more robust and trustworthy problem-solving models. These systems are becoming more and more popular due to their ability to solve a wide range of complex real-world problems related to such aspects as imprecision, uncertainty, or high dimensionality among others. Accordingly, the HAISs approach to ID has been widely covered in recent years combining different AI techniques and paradigms such as:

  - Fuzzy logic, data mining (association rules and frequent episodes), and genetic algorithms [236].
  - Expert rules, k-means algorithm, and Learning Vector Quantization [237].
  - ANNs and fuzzy logic [238].
  - Genetic algorithms and fuzzy logic [239].
  - Genetic algorithms and the K-Nearest Neighbor (K-NN) algorithm [240].
  - ANNs, Support Vector Machines, and Multivariate Adaptive Regression Splines [241].
  - K-NN and ANNs [242].
  - Neuro-fuzzy networks, fuzzy inference systems, and genetic algorithms [243].
  - Genetic algorithms and fuzzy logic [244].
  - Decision trees and Support Vector Machines [245].
  - ANNs, fuzzy logic, and genetic algorithms [246].
  - Flexible neural trees, genetic algorithms, and particle swarm optimisation [247].
  - Support Vector Machines, Self-Organized Feature Maps, and genetic algorithms [248].
  - It is worth emphasizing the intelligent agents approach. As described in section 3.3, some different AI techniques have been incorporated into software agents to make these agents more intelligent.

  Differentiating itself from previous work, the main hybrid system designed in this work is entitled MOVICAB-IDS. It combines AI techniques for visualisation-based ID.

- **Inclusion of deliberative agents in a MAS for ID**

  Some complex agents (CBR-BDI) incorporating different intelligent collaborative techniques are included in the proposed MAS. Although mobile agents can provide an IDS with some advantages (mobility, overcoming network latency, robustness, and fault tolerance), they were not included in the designed MAS as some problems have not been completely overcome yet [214]: speed, volume of the code required to implement a mobile agent, deployment, limited methodologies and tools, security threats, and so on.

- **A test method for visualisation-based IDSs**

  Previous work has presented few techniques to test and evaluate NIDSs as described in 6. In this study a novel testing technique is proposed to validate visualisation-based IDSs employing numerical datasets (see 6).

- **Application of unsupervised neural models for traffic data projection**

  Several attempts have been made to apply ANNs to the field of ID mainly based on a classificatory standpoint. A different approach is followed in this research line, in which the main goal is to provide the network administrator with a snapshot of the network traffic.

  Improving the effectiveness of visualisation for analysts at network/security operational centres might not require new visual idioms [174]. Accordingly, the existing combination of projection techniques and scatter plots already constitutes a very useful visualisation tool to investigate the intrinsic structure of multidimensional datasets, allowing experts to study the relations between different projections depending on the employed technique. This idea has been applied in the present study, where projection techniques support visual searches of structure (both normal and anomalous) in network-traffic datasets.

  Some other visualisation-based IDSs work at packet level, but most of them do not provide an intuitive visualisation of packets by preserving the temporal context. Thus, previous training is required to identify the "visual fingerprints" of attacks. This problem is overcome in MOVICAB-IDS by combining the selected neural projection model and the scatter plot technique. As a result of this combination, the temporal context of data (packets) is preserved in the views and then a synthetic and intuitive view of traffic is provided.

  Some other visualisation techniques could have been chosen but they were discarded due to their limitations. Such is the case of parallel coordinate plots, which, as a simple complex technique, can represent a high number of dimensions (up to 25 according to academic researchers [45]). Unfortunately, parallel coordinate plots suffer from occlusion, which is due to overlapping data that can leave both the line segments and the labels illegible. Additionally, parallel coordinate plots can not accurately display the quantity of visualised packets.

  From a purely "projection-of-packets" standpoint, some dimensionality reduction techniques such as PCA (described in section 2.5.6), have previously been proposed for visualising network data through scatter plots:

  - The PCA-based visualisation of the DARPA dataset provided in [249] does not support a clear distinction of attacks from normal traffic. Furthermore, no explanation of the projection obtained by this technique is provided.

- In [250, 251], PCA is proposed as a complementary tool to interpret the results obtained by a statistical analysis. Employing 12 packet features, this PCA visualisation does not by itself allow the identification of attacks. These authors extend their previous work in [151] by applying PCA, together with some other exploratory multivariate analysis methods and visualisation techniques, to network security. The k-means algorithm, hierarchical clustering, self-organizing maps, PCA, Independent Component Analysis (ICA), star plots, and mosaic plots are applied to the DARPA 1998 dataset. As pointed out by authors for the PCA-based projection: "*the distinction between normal and attack traffic is not clear*". Moreover, for the other projection-based visualisation (ICA), they also report that it is "*similar to the results from PCA... there is no clear distinction between normal and attack traffic*". According to them, the main reason for that is the high dependency between the columns of the dataset.
- [252] proposes the "*grand tour*" in the GGobi [253] visualisation tool for ID. Although some attacks are visually identified by combining visualisation and fuzzy feature extraction, no explanation of the projection technique and the visual ID process is provided.

PCA has been also applied [254, 255, 256, 257, 258, 259, 260, 261, 262] for dimensionality reduction as an initial step for further processing such as clustering, time-series analysis, or outlier detection.

The experimental outcomes of this work (see 5) outperform the above-mentioned results, because the CMLHL projections used in this work are based on higher-order statistics. The novel IDS presented in this book builds on previous research; the detection performance is improved and a fuller explanation of the use and the results obtained with these projection methods is provided.

- **Target attacks**

  Probing and some other very common and simple attacks have been addressed by many previous works on visualisation-based ID. However, little effort has been devoted to other more-specific attacks. MOVICAB-IDS is proposed for the detection of SNMP related anomalous situations (described in section 4.1) and some general probing techniques. Other authors [172] have also focused on these attacks by analysing network packets, although further details on the analysed attacks are not provided. They monitor the activity in the SNMP default port number (161). MOVICAB-IDS does not rely on the port number hosting the SNMP service because (as is recommended) it may well change.

This chapter has presented the state-of-the-art network-based ID from the visualisation and agent–based standpoints. It has compared the proposed hybrid IDS with previous work and has emphasised its main novelties. A further discussion on the features, advantages, and results of the hybrid IDS can be found in Chapter 7.

Chapter 4
A Novel Hybrid IDS

This book advances previous work by proposing **MOVICAB-IDS** (**MO**bile **VI**sualisation **C**onnectionist **A**gent-**B**ased **IDS**) [263] as a MAS-based IDS for network traffic visualisation. Network monitoring and ID can be easily performed by visualising packets travelling along the network through a neural projection model. So, the main goal of this hybrid IDS, which combines different AI paradigms, is to detect anomalous situations taking place in a computer network by means of visualisation.

MOVICAB-IDS supports a continuous inspection of network progress by analysing individual packets, which are intercepted by certain network capture devices. In this case, ID succeeds when significant deviations from normality are represented consistently in a visual manner in execution (virtually real) time.

Some of the ideas in previous works on hybrid AI IDSs (such as providing intelligence to agents, combining symbolic and subsymbolic paradigms, and others) which are described in this chapter are noted and incorporated in the MOVICAB-IDS formulation.

4.1 Target Attacks

In a computer network context, a protocol is a specification that describes low-level details of host-to-host interfaces or high-level exchanges between application programs. Among all the implemented network protocols, several can be considered highly dangerous in terms of network security.

Such is the case of SNMP (Simple Network Management Protocol) [264, 265], which was ranked as one of the top five most vulnerable services by CISCO [266]. Specially the two first versions [264, 267] of this protocol that still are the most widely used at present time. SNMP attacks were also listed by the SANS Institute as one of the top 10 most critical internet security threats [268, 269].

SNMP was oriented to manage nodes in the Internet community [264]. It is an application layer protocol that supports the exchange of management information (operating system, version, routing tables, default TTL, and so on) between network devices. This protocol enables network administrators to manage network performance and is used to control network elements such as routers, bridges, and

Á. Herrero and E. Corchado: Mobile Hybrid Intrusion Detection, SCI 334, pp. 71 – 89.
springerlink.com

switches. As a result, SNMP data are quite sensitive and liable to potential attacks. Indeed, an attack based on this protocol may severely compromise system security [270].

SNMP defines a simple set of operations (and the information these operations gather) that gives administrators the ability to change the state of any SNMP-based device [271]. SNMP can be used to shut down an interface on a router, to check the speed at which an Ethernet interface is operating, or to monitor the temperature on a switch. SNMP's predecessor, the Simple Gateway Management Protocol (SGMP), was developed to manage only Internet routers, but, on the contrary, SNMP can be used to manage other types of devices such as Unix and Windows systems, printers, modem racks, etc. Any device running software that allows the retrieval of SNMP information can be managed. This includes not only hardware but also software systems such as web servers and databases.

To date, several versions of SNMP [272] have been defined through standard RFCs by the IETF:

- SNMP version 1 (SNMPv1): up until a few years ago this was the standard version of SNMP. It was defined as a full IETF standard in RFC 1157 [264]. The only security mechanism it incorporates is communities, which are simply passwords: plain-text strings that allow an SNMP-based application knowing the community name to gain access to a device's management information. Three types of communities are defined in SNMPv1: read-only, read-write, and trap.
- SNMP version 2: is often referred to as community string-based SNMPv2. It is defined in RFC 3416, RFC 3417, and RFC 3418, which make RFC 1905, RFC 1906, and RFC 1907 obsolete, respectively. Some vendors started supporting it in practice and it is currently installed on many networks.
- SNMP version 3 (SNMPv3): this latest version of the protocol is defined in RFC 2571, RFC 2573, RFC 2574, RFC 3412, RFC 3414, RFC 3415, RFC 3416, RFC 3417, and RFC 3418. Other RFCs about SNMPv3 (RFC 3410 and RFC 3826) as yet they are not standard status. This version incorporates additional security mechanisms [273]: support for strong authentication and private communication between managed entities.

Some other non-standard RFCs are devoted to SNMP issues such as applications making use of an SNMP engine (RFC 3413) and the coexistence between the different versions (RFC 3584).

Two kinds of entities are defined in SNMP as shown in Fig. 4.1: agents and managers (also referred as Network Management Stations - NMSs). An SNMP agent is the operational role assumed by an SNMP party (generally a device controlled by this protocol) when it performs SNMP management operations in response to received SNMP messages. An SNMP proxy agent is an SNMP agent that performs management operations by communicating with another logically remote party. In the case of a segmented network, "logically remote" means that each party is located in a different network segment.

An SNMP manager can be defined as a server running a software system that can handle management tasks for a network. They are responsible for polling and receiving traps from SNMP agents in the network and consequently acting. Traps are messages asynchronously sent by SNMP agents, to tell managers that something has happened.

The transparency principle [265] states that the manner in which one SNMP party processes SNMP messages received from another SNMP party should be entirely transparent to the latter. Implicit in this principle is the requirement that, while interacting with a proxy agent, an SNMP manager should never be supplied with information on the nature or progress of the proxy mechanisms by which its requests are realized. Hence, it should appear to the manager to be directly interacting with the proxied device via SNMP.

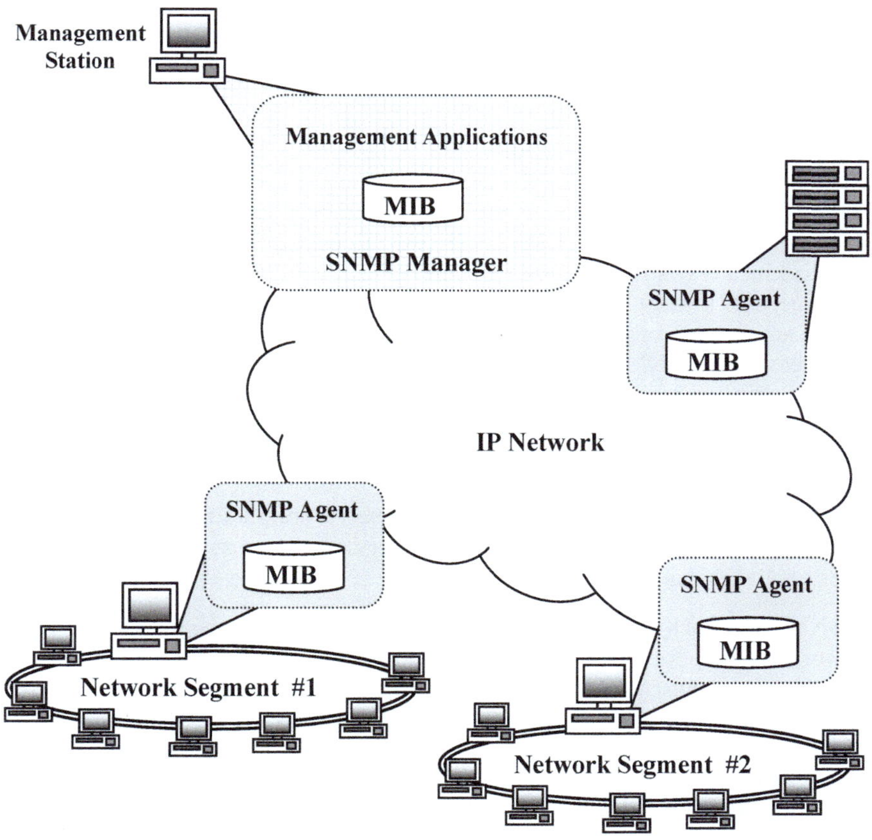

Fig. 4.1 SNMP-managed configuration (adapted from [274]).

SGMP (the predecessor to SNMP) set the use of a well-defined subset of the Abstract Syntax Notation One (ASN.1) language. SNMP continues this idea by using a moderately more complex subset of ASN.1 for describing managed objects and the protocol data units used for managing those objects.

The Management Information Base (MIB) can be defined in broad terms as the database used by SNMP to store information about the elements that it controls. Like a dictionary, an MIB defines a textual name for a managed object and explains its meaning. An agent may implement many MIBs, including features defined by the manufacturer of the device (referred to as a proprietary MIB), but all agents implement a standard MIB called MIB-II [275]. This standard defines variables for things such as interface statistics (speeds, maximum transmission unit, octets sent, octets received, etc.), as well as various other items pertaining to the system itself (system location, system contact, etc.). The main goal of MIB-II is to provide general TCP/IP management information [271].

The Internet Activities Board (IAB) recommended that all IP and TCP implementations were network manageable [276]. The implementation of the MIB and at least one of the management protocols such as SNMP is the consequence of this suggestion.

4.1.1 SNMP Attacks

Three types of computer attacks are most commonly reported by IDSs: system scanning, denial of service (DOS), and system penetration. These attacks can be launched locally (on the attacked machine) or remotely (using a network to access the target) [17]. Most security tools focus their attention on external attacks but attacks are just as likely to come from inside the network as from the outside. This is the reason why SNMP attacks must be considered; SNMP traffic can (and must) be easily rejected by a firewall but only if it comes from the outside.

As previously stated, consideration should be given to SNMP from a security standpoint, due to its very limited security mechanisms and the security sensitive data that is stored in the MIB. Attackers can exploit these vulnerabilities in the SNMP for network reconnaissance and remote reconfiguration or shut down of SNMP devices. Thus, MOVICAB-IDS focuses on the most commonly reported types of attacks that target SNMP:

- **SNMP network scan:** three types of scans (or sweeps) have been defined: network scans, port scans, and their hybrid block scans [277]. Unlike other attacks, scans must use a real source IP address, because the results of the scan (open ports or responding IP addresses) must be returned to the attacker [171].

  A port scan (or sweep) may be defined as series of messages sent to different port numbers to gain information on its activity status. These messages can be sent by an external agent attempting to access a host to find out more about the network services that this host is providing. So, a scan is an attempt to count the services running on a machine (or a set of machines) by probing each port for a response, providing information on where to probe for weaknesses. Thus, scanning generally precedes any further intrusive activity. This work focuses on the identification of network scans, in which the same port (the SNMP port) is the target for a number of computers in an IP address range. A network scan is one of the most common techniques used to identify services that might then be accessed without permission [141].

- **SNMP community search:** this attack is generated by a hacker sending SNMP queries to the same port number of different hosts, and by using different strategies (brute force, dictionary, etc.) to guess the SNMP community string. Once the community string has been obtained, all the information stored in the MIB is available for the intruder. The unencrypted community string can be seen as the SNMP password for versions 1 and 2. In fact, it is the only SNMP authentication mechanism.

  As reported in [268], "*The default community string used by the vast majority of SNMP devices is "public", with a few clever network equipment vendors changing the string to "private" for more sensitive information*". This eases SNMP intrusions, as intruders can guess the community string with little effort.

- **MIB information transfer:** this situation involves a transfer of some (or all the) information contained in the SNMP MIB, generally through the *Get* command or similar primitives such as *GetBulk* [278, 279]. This kind of transfer is potentially quite a dangerous situation because anybody possessing some free tools, some basic SNMP knowledge and the community string (in SNMP versions 1 and 2), will be able to access all sorts of interesting and sometimes useful information. As specified by the Internet Activities Board, the SNMP is used to access MIB objects. Thus, protecting a network from malicious MIB information transfer is crucial. However, the "normal" behaviour of a network may include queries to the MIB. This is a situation in which visualisation-based IDSs are quite useful; these situations may be visualised as anomalous by an IDS but it is the responsibility of the network administrator to decide whether or not it constitutes an intrusion.

The main reason for focusing on such attacks is that all these situations can be very risky on their own and all together: a network scan followed by an SNMP community search and ending with an MIB information transfer constitute an SNMP attack from scratch. An intruder gets some of the SNMP-managed information without having any previous knowledge about the network being attacked.

4.2 System Overview

The novel IDS presented in this study is based on the application of different AI paradigms to process the continuous data flow of network traffic. In order to do so, MOVICAB-IDS splits massive traffic data into limited datasets and visualises them, thereby providing security personnel with an intuitive snapshot to monitor the events taking place in the observed computer network.

The following AI paradigms are combined within MOVICAB-IDS:

- **Multiagent system:** some of the components are wrapped as deliberative agents capable of learning and evolving with the environment [280].
- **Case-based reasoning:** some of the agents contained in the MAS are known as CBR-BDI agents [281] because they integrate the BDI (Beliefs, Desires and Intentions) [282] model and the CBR (Case-Based Reasoning) paradigm.
- **Artificial neural networks:** the connectionist approach fits the intrusion-detection problem mainly because it allows a system to learn, in an empirical

way, the input-output relationship between traffic data and its subsequent interpretation [283]. The previously described CBR-BDI agents incorporate the CMLHL neural model (described in section 2.5.12) to generate projections of network traffic.

The combination of these paradigms allow the user to benefit from certain properties of ANN (generalization that allows the identification of previously unseen attacks), CBR (learning from past experiences), and agents (reactivity, proactivity, sociability, and intelligence), which greatly facilitates the ID task.

As proposed for traffic management [284], different tasks perform traffic monitoring and ID. For the data collecting task, a 4-stage framework [285] is adapted to MOVICAB-IDS in the following way:

1. **Data capture**

   As network-based ID is pursued, the continual data flow of network traffic must be managed. This data flow contains information on all the packets travelling along the network to be monitored. This stage is comprehensively described in section 4.2.1.

2. **Data selection**

   NIDSs have to deal with the practical problem of high volumes of quite diverse data [286]. To manage high diversity of data, MOVICAB-IDS splits the traffic into different groups, taking into account the protocol (UDP, TCP, ICMP, and so on) over IP, as there are differences between the headers of these protocols. Once the captured data is classified by the protocol, it can be processed in different ways. This stage is comprehensively described in section 4.2.1.

3. **Segmentation**

   The two first stages do not deal with the problem of continuity in network traffic data. The CMLHL model (as some other neural models) can not process data "on the fly". To overcome this shortcoming, a way of temporarily creating limited datasets from this continuous data flow is proposed by segmentation, as is described in section 4.2.2.

4. **Data pre-processing**

   Finally, the different datasets (simple and accumulated segments) must be preprocessed before presenting them to the neural model. At this stage, categorical features are converted into numerical ones. This happens with the protocol information; each packet is assigned a previously defined value according to the protocol to which it belongs.

Once the data-collecting task is performed and the data is ready, the MOVICAB-IDS process performs two further tasks:

- **Data analysis:** CMLHL is applied to analyse the data. Some other unsupervised models have also been applied to perform this task for comparison purposes. This task is described in section 4.2.3 and the results can be found in 7.

- **Visualisation:** the projections of simple and accumulated segments are presented to the network administrator for scrutiny and monitoring. One interesting feature of the proposed IDS is its mobility; this visualisation task may be performed on a different device other than the one used for the previous tasks. To improve the accessibility of the system, results may be visualised on a mobile device (such as phones or blackberries), enabling informed decisions to be taken anywhere and at any time. This task is described in section 4.2.4.

In summary, the MOVICAB-IDS task organisation comprises the six tasks depicted in Fig. 4.2.

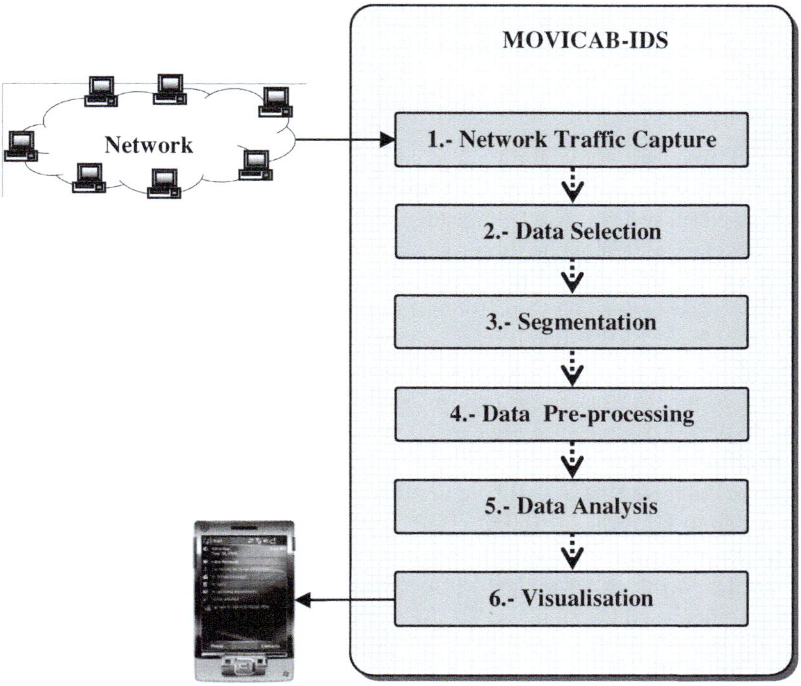

Fig. 4.2 MOVICAB-IDS task overview.

4.2.1 Network Traffic Capture and Selection

The ID process starts by capturing packets travelling along the network by means of sniffing techniques. One of the network interfaces can be set up in promiscuous mode to gather all the information travelling along the network. Every single packet is then captured and its header information is stored. As packets are wrapped according to the different protocols and applications in the TCP/IP protocols stack, some fields for the widely used TCP/IP protocols and applications are:

- TTL (Time to Live): timer used to track the lifetime of the packet.
- TOS (Type of Service): parameters for the type of service requested.
- Protocol: a code identifying the next encapsulated protocol. In the case of IP header, this field identifies the protocol over IP.
- Source IP address: IP address from where the source host sent the packet.
- Destination IP address: IP address to which the packet is sent.
- Source port: port number from where the source host sent the packet.
- Destination port: destination host port number to which the packet is sent.
- Acknowledgment information (control bit and number): in the case of TCP, the reception of packets is acknowledged back to the sender.
- Size: total packet size in bytes.
- Timestamp: the time when the packet was sent.

Widely-used freeware tools for packet sniffing (such as Wireshark [287]) are available for network traffic capture. Additionally, some other ways of collecting data such as Netflow [288] can be considered as an alternative for the data capture stage. In this case, data must be pre-processed in a different way.

Several approaches deal with traffic data that are summarized in TCP connections, such as the well-known KDD Cup 1999 dataset [153, 154]. On the contrary, MOVICAB-IDS employs a packet-based approach, where each instance in the final datasets to be visualised corresponds to a single packet, and only several features of each packet are used for ID. This means that MOVICAB-IDS is unable to detect intrusions that relate to packet payloads; an issue that is further discussed in Chapter 7.

As previously stated, to deal with the practical problem of high volumes of quite diverse data, MOVICAB-IDS splits the traffic into different categories, taking into account the protocol (UDP, TCP, ICMP, and so on) over IP. As SNMP attacks are targeted and the SNMP is based on UDP, greater consideration is given to UDP traffic in the experimental setup of this work. Additionally, TCP traffic has been also analysed through MOVICAB-IDS as described in section 5.2.

Once the traffic is split according to the protocol, a reduced set of features contained in the headers of the captured packets is selected. In the data selection stage, the following five variables of each packet are extracted: source port, destination port, size, timestamp, and protocol over TCP/UDP. As is shown in the experimental section of this work (Chapter 5), these five features allow the identification of the previously described SNMP attacks. Furthermore, this minimal traffic measurement [254], characterizing network packets by a reduced set of packet header features, allows the IDS to monitor high-volume networks.

4.2.2 Data Segmentation

To process the continuous flow of network traffic, MOVICAB-IDS splits the pre-processed data stream into simple and accumulated segments as depicted in Fig. 4.3. These segments are defined as follows:

- **Equal simple segments (S_x):** each simple segment contains all the packets with timestamps between the initial and final time limits of the segment. There must be a time overlap between each pair of consecutive simple segments because anomalous situations could conceivably take place between simple segment S_x and S_{x+1} (where S_{x+1} is the next segment following S_x).
- **Accumulated segments (A_x):** each one of these segments contains several consecutive simple ones. To avoid duplicated packets, time overlap is removed in accumulated segments.

One of the main reasons for such a partitioning is to present a long-term picture of the evolution of network traffic to the network administrator, as it allows the visualisation of attacks lasting longer than the length of a simple segment. Additionally, simple segments allow an execution-time processing of traffic. The longer a dataset is, the further the visualisation will be from the execution-time due to capture and analysis delays. Furthermore, simple segments help to overcome the problem of overplotting in scatter plots.

To avoid confusion on the part of the analyst, accumulated segments are visualised at the same time. This will prompt the network administrator to realise that there is only one anomalous situation being visualised twice.

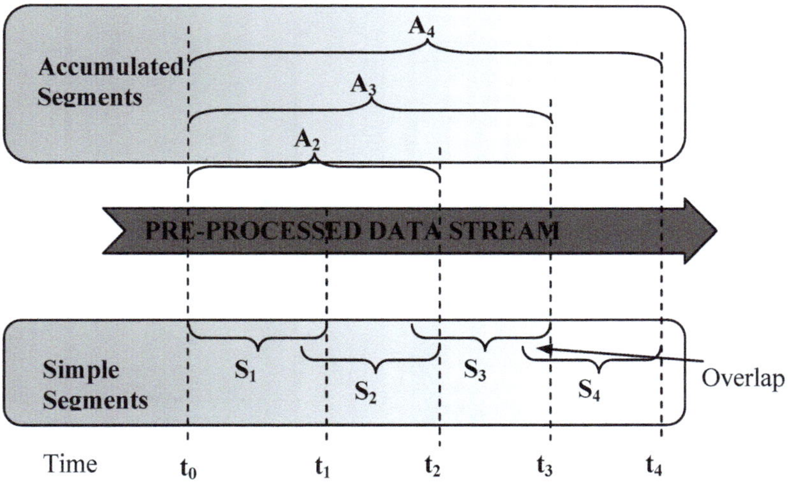

Fig. 4.3 MOVICAB-IDS segmentation of pre-processed data.

There are some key issues concerning segmentation which, among others, are: the length (time duration) of the simple segments, the overlap time, and the number of simple segments making up the accumulated segments. All of them must be defined after taking into account certain network features that need to be monitored such as the volume of traffic, rush hours, and so on.

4.2.3 Data Analysis

Once simple and accumulated segments have been built, a connectionist model will be applied to analyse them. The data analysis task is based on the use of CMLHL to drive a compact 2D or 3D visualisation of the 5-dimensional packet data. CMLHL is able to provide a projection showing the internal structure of a dataset by considering the fourth-order statistics (kurtosis index). Within MOVICAB-IDS, the proposed unsupervised neural projection technique operates on a feature-space representation of unlabeled network traffic.

4.2.4 Visualisation

Finally, the MOVICAB-IDS projections of the segments (both simple and accumulated) are displayed to the security personnel. The results are accessible from any wireless mobile device, such as a laptop, blackberry or mobile phone (see Fig. 4.4), which gives greater accessibility to network administrators, enabling permanent mobile visualisation, monitoring, and supervision of networks. Low-size static images are sent to these mobile devices due to their reduced visualisation capabilities.

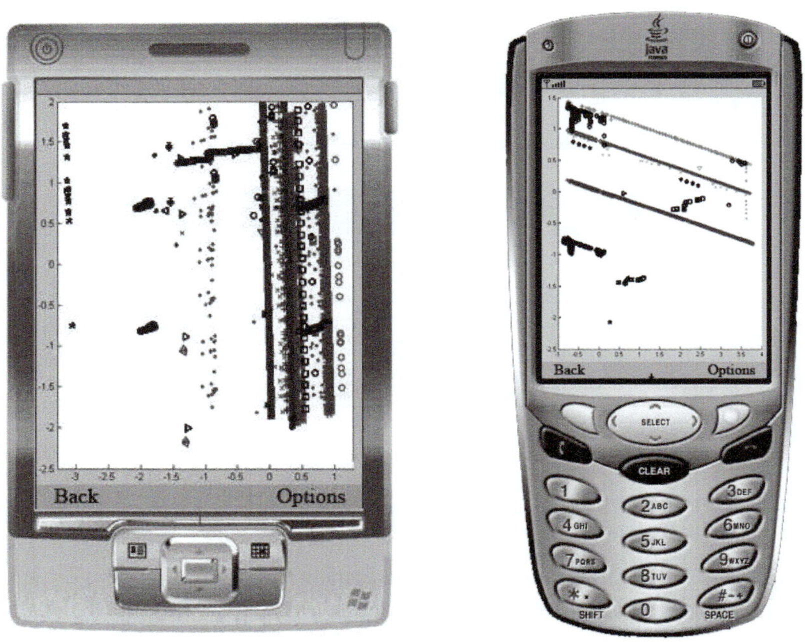

Fig. 4.4 MOVICAB-IDS sample visualisations.

4.3 Multiagent System

Having defined the tasks to be performed by MOVICAB-IDS, the extent to which this ID could incorporate most of the suggested characteristics for an agent-based solution (see section 2.6) was verified. Thus, it was decided to design MOVICAB-IDS as a MAS which contains both reactive and deliberative-type agents. This section briefly describes the process of MOVICAB-IDS development and the resulting MAS.

CBR-BDI agents [281, 289, 290] integrate the Case-Based Reasoning (CBR) paradigm and the Belief-Desire-Intention (BDI) architecture (both described in Chapter 1). By using CBR as a reasoning mechanism, these agents learn from initial knowledge, have a large capacity for adaptation to the needs of their surroundings, and interact autonomously with the environment, users, and other agents within the system. CBR-BDI agents provide planning based on previous experiences as CBR systems use memories (past experiences) to solve new problems. In the case of MOVICAB-IDS, they are applied to tune the unsupervised neural model called CMLHL.

An evolution of CBR-BDI agents, case-based planning - BDI (CBP-BDI) agents [291], have already been used in this research. In keeping with the main idea behind CBR, CBP [292] is based on solving new planning problems by reusing past successful plans [293]. One of the key points in the CBP-BDI based planning is the notation used to represent the solution (plans). A solution can be seen as a sequence of intermediate states transited to go from an initial state to the final state. States are usually represented as propositional logic sets. The set of actions can be represented as a set of operators together with an order relationship. In the case of MOVICAB-IDS, they are applied to optimise the scheduling of data analysis tasks.

4.3.1 Methodology

An extended version of the Gaia methodology [294] has been applied to design the MOVICAB-IDS MAS. This section shows the main outputs of the application of such methodology.

At the initial architectural design stage, the following roles were identified:

- Sniffer: continuously selects and captures the traffic data flowing along a network segment. At the same time, when sufficient captured data exist, these data are split and their readiness is communicated to other roles.
- Pre-processor: pre-processes the captured data and then requests an analysis for this new piece of data.
- Analyzer: negotiates for data analysis with other ANALYZER agents and with the COORDINATOR agent. Once an analysis is assigned, it analyses the new pre-processed data.
- ConfigurationManager: manages the configuration of several parameters (related to the splitting, pre-processing, and the analysis of traffic data) and makes the information available to some other roles.
- Coordinator: coordinates some of the other roles and balances the workload between them.

- Visualizer: responsible for updating the visualisation when new information (analysed data or system information) is generated.

The following protocols have been also defined at this stage:

- AnalysisAborted.
- AnalysisCompleted.
- ChangeSplitConfig.
- ChangePre-processConfig.
- ChangeAnalysisConfig.
- ManageSplitError.
- NegotiateAnalysis.
- Pre-processAborted.
- Pre-processedDataReady.
- RequestAnalysisConfig.
- RequestAnalysedData.
- RequestPre-processConfig.
- RequestPre-processedData.
- RequestSplitConfig.
- RequestSplitData.
- RequestVisualisation.
- SplitAborted.
- SplitDataReady.
- UpdateAnalysisConfig.
- UpdatePre-processConfig.
- UpdateSplitConfig.
- UpdateSystemInfo.

The detailed design stage concluded that there is a one-to-one correspondence between roles and agent classes in the system, as can be seen in the agent model (Fig. 4.5).

$$\text{SNIFFER}^{1\ldots N} \xrightarrow{plays} \text{SNIFFER}$$

$$\text{PRE-PROCESSOR}^{+} \xrightarrow{plays} \text{PRE-PROCESSOR}$$

$$\text{ANALYZER}^{+} \xrightarrow{plays} \text{ANALYZER}$$

$$\text{CONFIGURATIONMANAGER}^{1} \xrightarrow{plays} \text{CONFIGURATIONMANAGER}$$

$$\text{COORDINATOR}^{1} \xrightarrow{plays} \text{COORDINATOR}$$

$$\text{VISUALIZER}^{+} \xrightarrow{plays} \text{VISUALIZER}$$

Fig. 4.5 MOVICAB-IDS agent model.

As a result, six agents (SNIFFER, PRE-PROCESSOR, ANALYZER, CONFIGURATIONMANAGER, COORDINATOR, and VISUALIZER) are included in the MOVICAB-IDS agent environment as can be seen in Fig. 4.6. They all are described in the following sections. Special attention is paid to the ANALYZER hybrid CBR-BDI agent, as it is the most complex one.

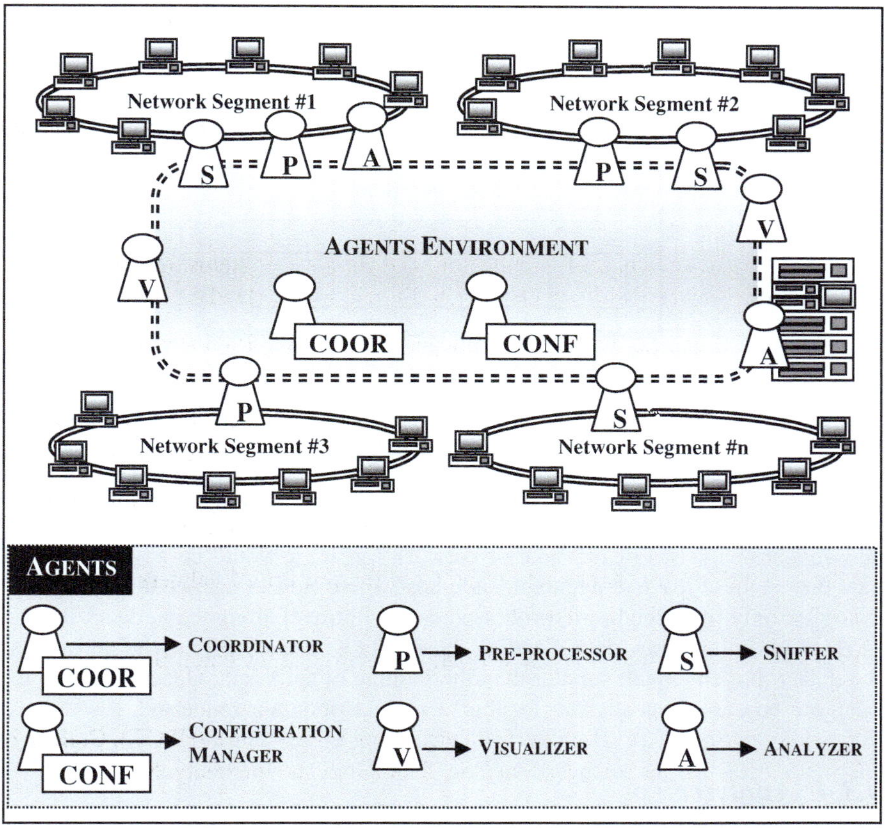

Fig. 4.6 MOVICAB-IDS agent environment.

4.3.2 *Sniffer*

This reactive agent is in charge of gathering, selecting, and segmenting the traffic data (see Fig. 4.7). The continuous traffic flow is captured and split into segments (as described in section 4.2.2) in order to send it through the network for further process and analysis. The readiness of the segmented data is communicated to the Pre-processor agent. One agent of this class is located in each of the network segments that MOVICAB-IDS has to monitor (from 1 to n). Additionally, there are backup instances (one per network segment) ready to substitute the active ones if they fail or are attacked because these agents are the most critical ones. ID can not be performed if traffic data is not captured.

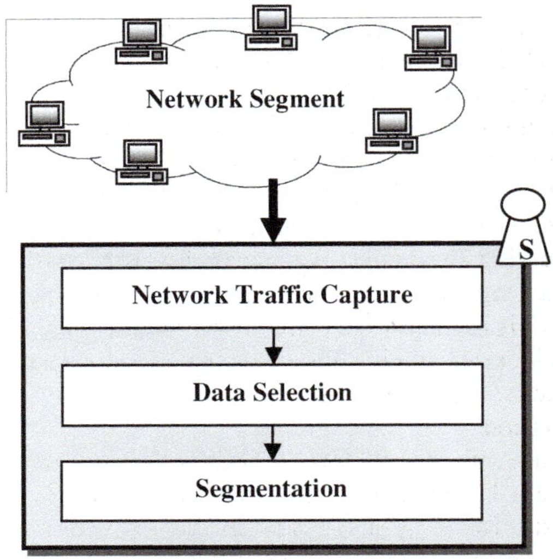

Fig. 4.7 Inner architecture of SNIFFER agents.

4.3.3 Pre-processor

After splitting traffic data, the generated segments must be pre-processed to apply subsequent analysis. For the sake of network traffic, it would be advisable to locate one of these reactive agents in each host where Sniffer agents are located. By doing so, only lightweight pre-processed data will travel along the network instead of the high-volume raw data. Sending these data will not overload the network as the data volume is much smaller than the volume of split data. Once the data have been pre-processed, an analysis for this new piece of data is requested.

4.3.4 Analyzer

This is a hybrid CBR-BDI deliberative agent. It employs CBR to tune the parameters of the CMLHL model (see section 2.5.12) to analyse pre-processed traffic data. This agent generates a solution (or achieves its goals) by retrieving a previously analysed case and analysing the new one through the CMLHL architecture. Cases are defined by several features, as can be seen in Table 4.1.

The Analyzer agent incorporates two different behaviours, namely "learning" and "exploitation". Initially, during the set-up stage, this agent incorporates new knowledge (modelled as sets of problem/solution) into the case base as previously described. This learning behaviour is characterized by the four below described CBR stages. Once the case base is wide enough, the exploitation behaviour is started. From then on, the revise and retain stages of the CBR cycle are no longer performed. When a new analysis request arrives, the Analyzer agent retrieves the

Table 4.1 ANALYZER agent - representation of case features. Classes: P (problem description attribute) and S (solution description attribute).

Class	Feature	Type	Description
P	Segment length	Integer	Total segment length (in ms).
P	Network segment	Integer	Network segment where the traffic comes from.
P	Date	Date	Date of capturing.
P	#source ports	Integer	Total number of source ports.
P	#destination ports	Integer	Total number of destination ports.
P	#protocols	Integer	Total number of protocols.
P	#packets	Integer	Total number of packets.
P	Protocol/packets	Array	An array (of variable length depending on each dataset) containing information about how many packets of each protocol there are in the dataset.
S	#iterations	Integer	Number of iterations.
S	Learning rate	Float	Learning rate.
S	p	Float	CMLHL parameter.
S	Lateral strength	Float	CMLHL parameter.
S	Weights	Matrix	A matrix containing the synaptic weights calculated by the CMLHL model after training.

most similar case previously stored in the case base. Then, the weights contained in the solution are reused to project the new data. The other parameters of the neural model are not reused, as the neural network is not trained again.

The techniques and tools used by the ANALYZER agent to implement the CBR stages (retrieval, reuse, revision, and retention) are depicted in Fig. 4.8 and defined as follows:

- **Retrieve:** when a new analysis is requested, the ANALYZER agent tries to find the most similar case to the new one in the database. Associative retrieval [293] based on Euclidean distance is used to find the most similar case in the multi-dimensional space defined by the main features characterising each case (see problem description features in Table 4.1).
- **Reuse:** once the most similar case has been selected, its solution is reused. This solution consists of the values of the parameters used to train the CMLHL model (see solution description features in Table 4.1).

  A set of trainings (for the CMLHL model with a combination of different parameter values varying in a specified range) is proposed by considering the distance between the new case and the most similar one. In other words, if they are very similar, a reduced set of trainings will be performed. On the contrary, if the most similar case is far away from the new one, a larger number of trainings with very different parameter values will be generated.
- **Revise:** the CMLHL model is trained for the new traffic segment using the combination of parameters values generated in the reuse stage. When the new

projections (the outputs of the CMLHL model for each parameter combination) of the segment are ready, they are shown to the human user through the VISUALIZER agent. The user then chooses one of these projections as the best one; the one that provides the clearest snapshot of the traffic evolution.

- **Retain:** once the best projection is selected by the user, the ANALYZER agent stores the new case containing the problem-descriptor and the solution (parameter values used to generate the chosen projection) in the case base for future reuse (see Table 4.1).

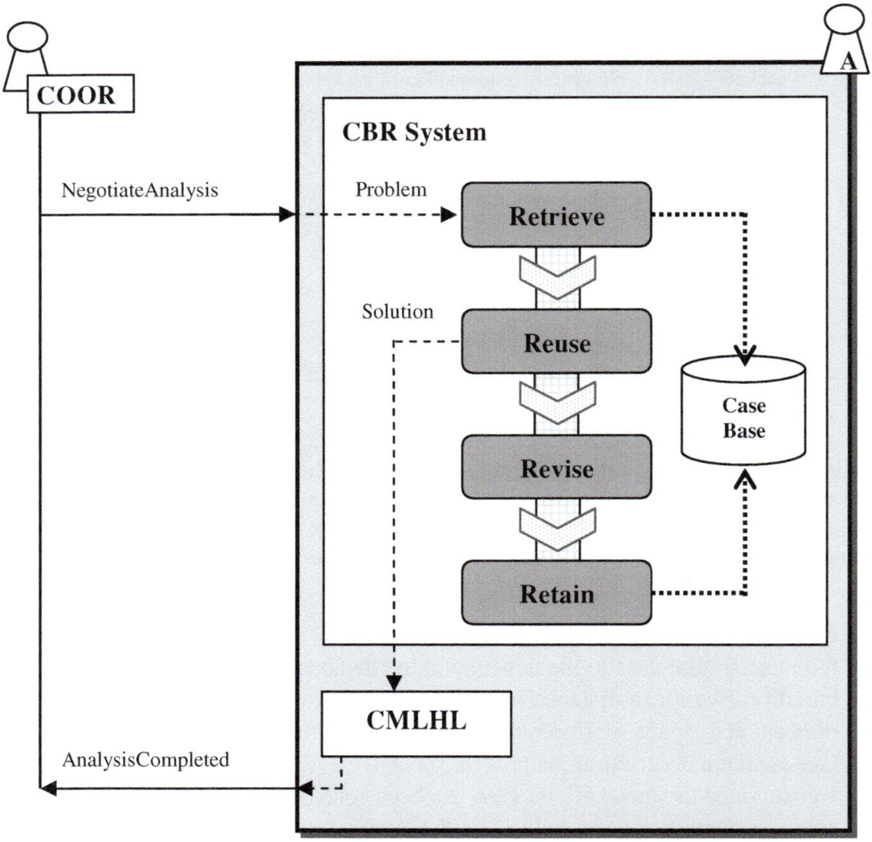

Fig. 4.8 Inner architecture of ANALYZER agents.

The ANALYZER is clearly the most resource-consuming class of agents as they train the neural model. The amount of computational resources needed to analyse the data coming from different network segments is extremely high. To overcome this demand, ANALYZER agents can be located in high-performance computing clusters or in less powerful machines whose computing resources are under-used (as can be seen in Fig. 4.6). In this way, MOVICAB-IDS can be adapted to the available resources for ID.

4.3.5 *ConfigurationManager*

It is worth mentioning the importance of the configuration information. The processes of data capture, split, pre-process, and analysis depends on the values of several parameters, as for example: packets to capture, segment length, or features to be extracted among others. All this information is managed by the CONFIGURATIONMANAGER reactive agent, which is in charge of providing this information under requests from the SNIFFER, PRE-PROCESSOR, and ANALYZER agents.

4.3.6 *Coordinator*

There is only one instance of the COORDINATOR agent, but there can be several ANALYZER agents (from 1 to m) located in a set of heterogeneous computers, following a cluster-like distribution. In order to improve system efficiency and perform execution-time processing, the pre-processed data must be dynamically and optimally assigned for analysis. The COORDINATOR agent, in charge of allocating the pending analyses to the available ANALYZER agents, is defined as a Case-Based Planning (CBP) -BDI agent [291]. CBP [292] attempts to solve new planning problems by reusing past successful plans [293]. Load balancing and scheduling in a cluster built of heterogeneous computers (such as the one supporting MOVICAB-IDS analysis task) is an interesting matter [295]. Intelligent agents have been previously applied to load balancing; in [296], a MAS is in charge of optimally load-balancing work between different agents, based on purely local information. In MOVICAB-IDS, general information about the ANALYZER agents is available for the COORDINATOR agent. In [297], a negotiation strategy was chosen to allocate jobs to different resources by means of a MAS. Aspects such as price, performance, and quality of service are considered. Additionally, a CBR is used to predict how long a resource will take to execute a job. This can be guessed on the basis of the machine's performance in past cases of running similar jobs with similar resources.

The COORDINATOR agent plans to allocate an analysis to one of the available ANALYZER agents based on the following criteria:

- **Location.** ANALYZER agents located in the network segment where the VISUALIZER or PRE-PROCESSOR agents are placed would be prioritised.
- **Available resources** of the computer where each ANALYZER agent is running. The computing resources and their rate of use all have to be taken into account. Thus, the work load of the computers must be measured.
- **Analysis demands.** The amount and volume of data to be analysed are key issues to be considered.
- **ANALYSER agents behaviour.** As previously stated, these agents behave in a "learning" or "exploitation" way. Learning behaviour causes an ANALYZER agent to spend more time over an analysis than exploitation behaviour does.

As a computer network is an unstable environment, the availability of ANALYZER agents changes dynamically. As network links can stop working from time to time, the COORDINATOR agent must be able to re-allocate the analyses previously assigned to the ANALYZER agents located in the network segment that may be down at any one time. An adaptation algorithm was designed to allow dynamic replanning in execution time.

The four stages of the CBP cycle of the COORDINATOR agent are defined as follows:

- **(Plan) Retrieve:** when a new pre-processed dataset is ready, an analysis is requested to the COORDINATOR agent. Associative retrieval is followed by taking into account the case/plan description shown in Table 4.2.

Table 4.2 COORDINATOR agent - representation of case features. Classes: P (problem description attribute) and S (solution description attribute).

Class	Feature	Type	Description
P	#packets	Integer	Total number of packets contained in the dataset to be analysed.
P	Analyzers / location	Array	An array (of variable length depending on the number of available ANALYZER agents) indicating the network segment where the ANALYZER agent is located.
P	Analyzers / features	Array	An array (of variable length depending on the number of available ANALYZER agents) containing information about the resources, their availability, and pending tasks.
P	Analyzers / failures	Array	An array (of variable length depending on the number of available ANALYZER agents) containing information about the number of times each ANALYZER agent has stopped working in the recent past (execution failures).
S	Analyzers / plans	Array	An array (of variable length depending on the number of available ANALYZER agents) containing the analyses assigned to each ANALYZER agent.

- **Reuse:** the retrieved plan is adapted to the new planning problem. The only restriction is that the analyses running at that time (the results of which have not yet been reported) cannot be reassigned. The others (pending) can be reassigned in order to optimise overall performance.
- **Revise:** the plan revision consists of a two-fold analysis. On the one hand, planning failures are identified by finding under-exploited resources. As an example, the following hypothetical situation is identified as a planning failure: one of the ANALYZER agents is not busy performing an analysis while the other ones have a list of pending analyses. On the other hand, execution failures are detected when communication with ANALYZER agents has been interrupted. Information on these failures is stored in the case base (as shown in Table 4.2) for future consideration. When an execution failure is detected, the CBP cycle is run from the beginning, which renews the analysis request.
- **Retain:** when a plan is adopted, the COORDINATOR agent stores a new case containing the dataset-descriptor and the solution (see Table 4.2).

4.3.7 *Visualizer*

At the very end of the ID process, the data that has been analysed are presented to the security personnel by means of a functional and mobile visualisation through the VISUALIZER interface agent. To improve the accessibility of the system, segment projections may be visualised on a mobile device (as can be seen in Fig. 4.4), enabling informed decisions to be taken anywhere and at any time. Visualisation facilities will differ according to the platform where the information is finally displayed.

Instances of the VISUALIZER agent can be run on both computers and mobile devices (such as phones, blackberries, etc.). Certain restrictions are associated with the use of mobile platforms that have fewer resources than others. That is the case of the Java 2 Platform - Micro Edition (J2ME), on which VISUALIZER agents are run over mobile devices. These agents can only provide a reduced set of interface features. Thus, two kinds of VISUALIZER agents are defined:

- **Mobile VISUALIZER agents:** run on mobile platforms providing reduced interface facilities.
- **Advanced VISUALIZER agents:** run on platforms without interface limitations and thus providing advanced interface facilities.

Chapter 5
Experiments and Results

Experimental results obtained by MOVICAB-IDS when examining normal and anomalous traffic are presented in this chapter. CMLHL projections enable MOVICAB-IDS to identify anomalous situations because these situations tend not to resemble parallel and smooth directions (as normal situations do) and because of their high packet concentrations, among others. The figures in this chapter include illustrations that highlight these anomalous situations, which it should be noted are not generated by MOVICAB-IDS.

As previously described in section 2.5.12, CMLHL generates output dimensions (projections) that are combinations of the input dimensions. Thus, the axes of output projections, X and Y, do not necessarily relate to one single input dimension.

MOVICAB-IDS was designed to detect anomalous situations relating to the SNMP (justified in section 4.1). As there is no publicly available packet-level dataset containing such attacks, it was decided to create a new dataset named GICAP-IDS dataset. The main experimental study of MOVICAB-IDS (section 5.1) makes use of this dataset.

Among the targeted SNMP anomalous situations, only port/network scans are contained in publicly available datasets, such as the DARPA dataset [1, 199, 200]. This chapter includes the results of dealing with port scans, which are contained in the well-known DARPA dataset (section 5.2), to verify the ability of MOVICAB-IDS to deal with such attacks. Note that this is not a complete benchmark aimed at comparing the performance of MOVICAB-IDS with some other previous IDSs.

The results in this chapter are discussed in section 7.1.

5.1 GICAP-IDS Dataset

As mentioned above, the GICAP-IDS dataset was generated "made-to-measure" in a small-size (28 hosts) network where "normal" traffic was known in advance. In addition to the SNMP packets, the dataset contains traffic related to some other protocols, considered as "normal" traffic. As this network was isolated and protected from external attacks, "normal" traffic was known in advance and has been empirically tested. MOVICAB-IDS captured packets relating to 63 different

Á. Herrero and E. Corchado: Mobile Hybrid Intrusion Detection, SCI 334, pp. 91–103.
springerlink.com

protocols. Further details about the network in which the traffic was captured are unavailable due to the security policy.

In this section, some scatter-plots are provided, each of which depicts all the packets contained in the segment projection that is shown. MOVICAB-IDS plots the packets in different colours and shapes taking into account the protocol information. Thus, all the packets assigned to the same protocol are depicted in the same way, which, as explained in this section, leads to an intuitive visualisation of network traffic.

5.1.1 Dataset Description

Three main anomalous situations related to the SNMP are distributed throughout the different segments in this study, namely: network scans, SNMP community searches and MIB (Management Information Base) information transfers. These situations can imply security risks, as described in section 4.1.

By performing the SNMP community searches between the port/network scans and the MIB transfers, all these anomalous situations constitute an SNMP attack from the outset. After trying to find whether and where (hosts and port numbers) SNMP is running, the community string is guessed in order to access the information contained in the MIB. When the community string has been found, the MIB information is read.

The following subsections describe the anomalous situations in depth that are found in the GICAP-IDS dataset.

5.1.1.1 Network Scans

A port/network scan provides information on where to probe for weaknesses, for which reason scanning generally precedes any further intrusive activity [138]. A network scan is one of the most commonly used techniques to identify services that might be accessed without permission [141].

To check the generalization capability of the proposed IDS, port numbers that are different from the default SNMP ports (161 and 162) were scanned. The response of MOVICAB-IDS showed its ability to detect network scans aimed at any port number, including those aimed at SNMP ports.

The GICAP-IDS dataset contains network scans aimed at port numbers 1434 (registered port number) and 65788 (as an example of dynamic or private port). In other words, SNMP packets were sent to the port numbers of all hosts in the IP address range that was monitored. These network scans were performed with the SNMPing tool that receives the replies from the hosts in the scanned IP address range. It summarises the information that was gathered, stating whether there was an answer from the host, whether the SNMP was disabled or configured on another port, or whether the SNMP was running.

5.1.1.2 SNMP Community Search

The unencrypted "community string" can be seen as the only authentication mechanism of SNMP versions 1 and 2. An SNMP community search is characterised by the intruder sending SNMP queries to the same port number of different hosts trying to guess the SNMP community string by means of different strategies such as brute force or dictionary techniques [269]. Once the community string has been obtained, all the information stored in the MIB is available for the intruder.

In the GICAP-IDS dataset, the attacker probes for three different community names on three different port numbers. In other words, three different hosts try to check three different community strings ("public", "private", and "aab") for SNMP on ports 161, 1161 and 2161. "public" and "private" are the default community strings for most SNMP-configured network devices, while "aab" was included as a different test string. Port numbers 1161 and 2161 were included in these searches as the SNMP can be configured to a port number other than the default port numbers: 161 and 162. The SNMP community searches contained in the GICAP-IDS dataset were performed using the Solarwinds Network Management Tools.

5.1.1.3 MIB Information Transfer

This situation involves a transfer of some (or all) of the information contained in the SNMP MIB, which is potentially quite dangerous, as the MIB contains very important network configuration data.

The MIB information transfer included in the GICAP-IDS dataset was generated by the get-bulk command; i.e. all the information stored in the MIB was sent through the network. These MIB transfers were performed with specific developed software.

5.1.1.4 Segments Content

As is described in Chapter 4, traffic data are split in different segments by MOVICAB-IDS. In this experimental study, the simple segment length was set up to ten minutes and the overlapping time between consecutive simple segments was 2 minutes.

Table 5.1 describes the segments used in this study that were generated according to the above mentioned values. For the sake of brevity, only some of these segments were selected for including their visualisations in this chapter. The traffic contained in these selected segments can be roughly described as:

- **S_1:** Contains only "normal" traffic, that is, no anomalous situations are included in this segment. As a consequence, its projection provides a sample MOVICAB-IDS visualisation of how "normal" traffic behaves.
- **S_2:** Apart from "normal" traffic, it contains two network scans (anomalous situations) that target port numbers 1434 and 65788 of all the machines within an IP address range.
- **S_3:** Contains "normal" traffic and SNMP community searches that target port numbers 161, 1161, and 2161 of all the machines within an IP address range. Three different community names were used for each one of these port numbers. Due to the time overlap, the last packets of the network scans contained in S_2 are also included in S_3.

- **S_4:** Contains "normal" traffic and an MIB information transfer generated by the *get-bulk* SNMP command.
- **A_2:** As it is a compilation of the traffic contained in segments S_1 and S_2, this segment contains two network scans aimed at port numbers 1434 and 65788.
- **A_3:** In addition to the network scans contained in A_2 (aimed at port numbers 1434 and 65788), it also contains the SNMP community searches included in S_3.
- **A_{13}:** This is the longest segment projection that is shown in this chapter. It contains examples of all the anomalous situations previously described: network scans (port numbers 1434 and 65788), SNMP community searches (port numbers 161, 1161 and 2161), and two MIB information transfers.

Table 5.1 GICAP-IDS dataset - simple and accumulated segments description.

Dataset	Number of packets	Initial time limit (ms)	Final time limit (ms)
S_1	3,122	1	600,000
S_2	3,026	480,000	1,080,000
S_3	3,235	960,000	1,560,000
S_4	9,673	1,440,000	2,040,000
S_5	10,249	1,920,000	2,520,000
S_6	3,584	2,400,000	3,000,000
S_7	3,051	2,880,000	3,480,000
S_8	2,818	3,360,000	3,960,000
...			
A_2	5,553	1	1,080,000
A_3	8,219	1	1,560,000
A_4	17,262	1	2,040,000
A_5	20,410	1	2,520,000
A_6	23,352	1	3,000,000
A_7	25,633	1	3,480,000
A_8	27,970	1	3,960,000
...			
A_{13}	49,647	1	6,360,000

5.1.2 Results

5.1.2.1 Simple Segments

The following figures show some examples of how MOVICAB-IDS visualises simple segments of 10-minute length.

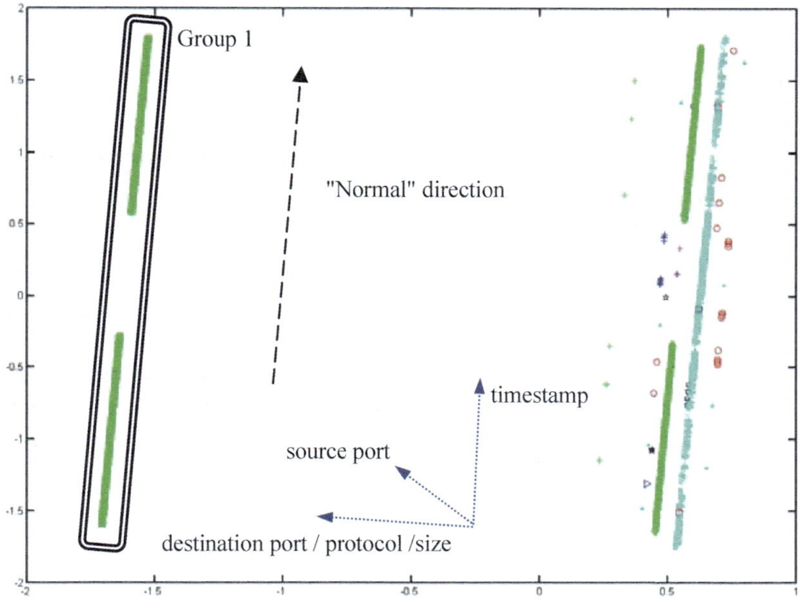

Fig. 5.1 MOVICAB-IDS visualisation of S_1 segment.

Table 5.2 Parameter values of Fig. 5.1.

Parameter values of Fig. 5.1	
Parameter	**Value**
Iterations	100,000
Learning rate	0.03
p	0.3
τ	0.12

Fig. 5.1 shows the projection of a simple segment (S1) containing no anomalous situations. This is the way that MOVICAB-IDS depicts "normal" traffic: parallel straight lines. After analysing each packet that is depicted, it was noticed that a certain ordering related to the input variables is preserved in this and other projections. The original dimensions of the dataset are preserved as indicated in Fig. 5.1.

Any sign of non-parallel evolution or high packet concentrations is viewed as an anomaly. It can be seen how in this figure all the packets (related to "normal" traffic) evolve in parallel "lines". For some protocols, we can not define a proper "line" because there are not enough packets. We can draw a line crossing all these packets (from the same protocol) in the plot. This line will be then parallel to the others.

Additionally, Fig. 5.1 allows us to identify a disruption in a protocol traffic. As can be seen in this figure, the traffic related to a certain protocol (Group 1 in

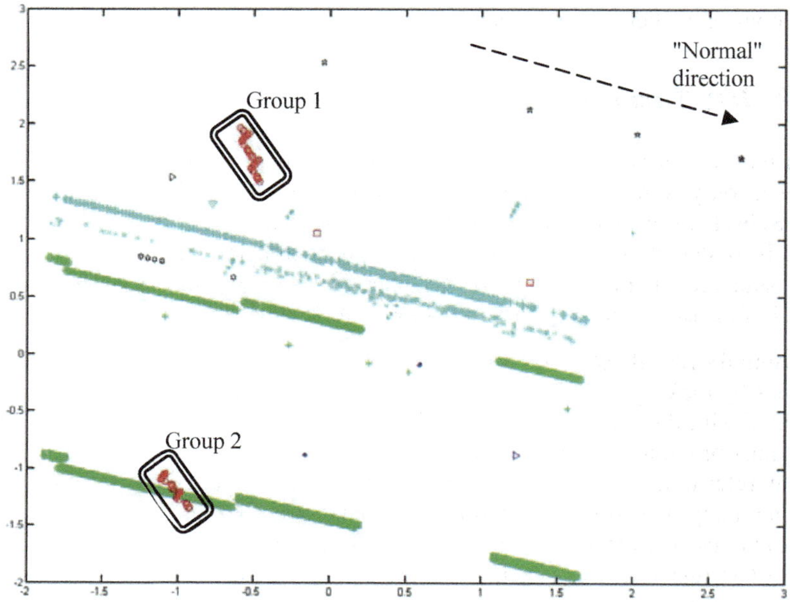

Fig. 5.2 MOVICAB-IDS visualisation of S_2 segment.

Table 5.3 Parameter values of Fig. 5.2.

Parameter values of Fig. 5.2	
Parameter	**Value**
Iterations	100,000
Learning rate	0.03
p	0.3
τ	0.12

Fig. 5.1) is interrupted at a certain point. Thus, the network administrator should realise that this protocol stopped working for a while. This requires an in-depth investigation to ascertain the reasons for such an interruption, as it might not be related to an intrusion.

MOVICAB-IDS visualisation of S2 segment is shown in Fig. 5.2. As in the previous segment, most of the traffic (identified as "normal") evolves in parallel straight lines. Additionally, there are two "groups" (Groups 1 and 2) of packets that are not depicted in the same way. Looking at the source data, it was checked that all these packets (visualised in a non-parallel line to normal traffic) formed part of the SNMP network scans contained in S2 segment. Packets contained in Group 1 were related to the network scan aimed at port number 1434, while packets contained in Group 2 made up the scan aimed at port number 65788.

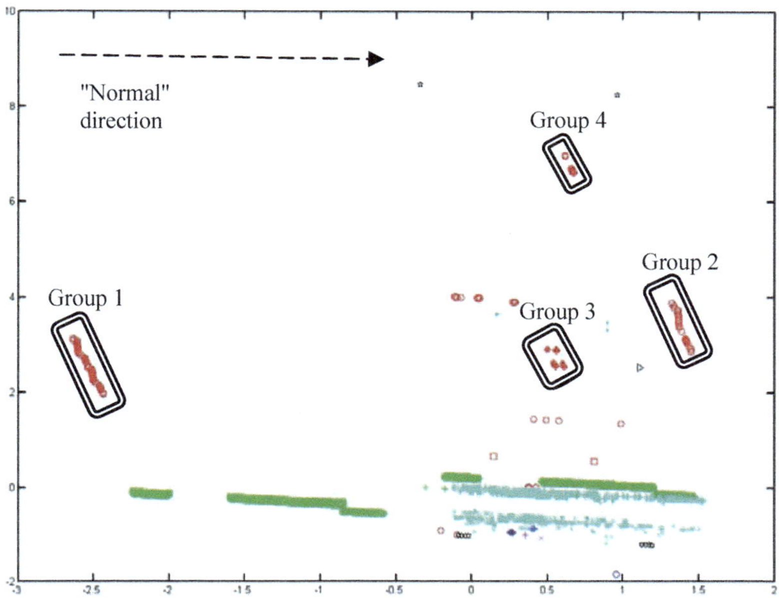

Fig. 5.3 MOVICAB-IDS visualisation of S_3 segment.

Table 5.4 Parameter values of Fig. 5.3.

Parameter values of Fig. 5.3	
Parameter	**Value**
Iterations	100,000
Learning rate	0.03
p	1.3
τ	0.12

As can be seen in Fig. 5.3, some anomalous groups can be identified in the visualisation of S3 segment. To begin with, Groups 1 and 2 contain the packets related to the network scans aimed at port numbers 1434 and 65788 respectively. Due to the time overlap, last packets of the network scans contained in S2 segment are also included at the beginning of S3 segment. As in the previous segment, these groups of packets evolve in a non-parallel direction to "normal" traffic.

Apart from Groups 1 and 2, the other two groups (Groups 3 and 4 in Fig. 5.3) show an anomalous evolution. These other anomalous groups are related to the SNMP community searches aimed at port numbers 161, 1161, and 2161. It is shown how MOVICAB-IDS visualises these intrusions (network scans and SNMP community searches) in the same fashion: lines that are non-parallel to "normal" traffic.

As displayed in Table 5.4, the p parameter was tuned in a different way for the S3 segment. This shows the way in which the MOVICAB-IDS Analyzer agent tunes the CMLHL parameters according to the nature of the analysed segments. It can be also seen in the parameter values for accumulated segments (see section 5.1.2.2), which are different from those for simple segments.

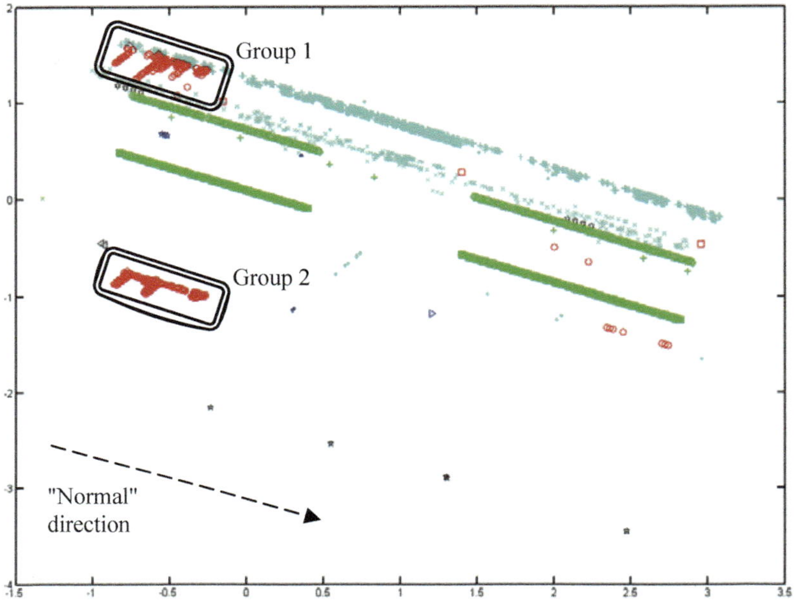

Fig. 5.4 MOVICAB-IDS visualisation of S_4 segment.

Table 5.5 Parameter values of Fig. 5.4.

Parameter values of Fig. 5.4	
Parameter	**Value**
Iterations	100,000
Learning rate	0.03
p	0.3
τ	0.12

Fig. 5.4 (visualisation of dataset S4) shows the way in which the system identifies an anomalous situation related to an MIB information transfer. This situation (Groups 1 and 2) is identified as anomalous due to its high packet concentrations (in comparison to the "normal" traffic) and once again, its evolution does not fit straight lines as the "normal" traffic does.

5.1.2.2 Accumulated Segments

The following figures show some examples of how MOVICAB-IDS visualises accumulated segments of variable length:

- an 18 minute-long accumulated segment (A_2 - Fig. 5.5).
- a 26 minute-long accumulated segment (A_3 - Fig. 5.6).
- a 106 minute-long accumulated segment (A_{13} - Fig. 5.7).

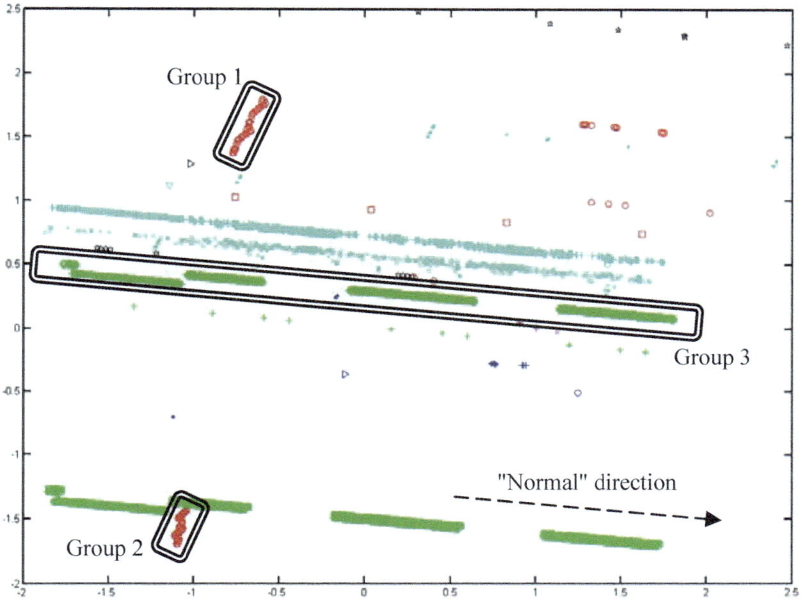

Fig. 5.5 MOVICAB-IDS visualisation of A_2 segment.

Table 5.6 Parameter values of Fig. 5.5.

Parameter values of Fig. 5.5	
Parameter	**Value**
Iterations	100,000
Learning rate	0.03
p	0.3
τ	0.12

As previously shown in Fig. 5.2, MOVICAB-IDS identifies network scans (Groups 1 and 2 in Fig. 5.5) due to their evolution that is non-parallel to "normal" traffic. Additionally, traffic disruptions in a protocol (Group 3 in Fig. 5.5) can also be identified as in Fig. 5.1.

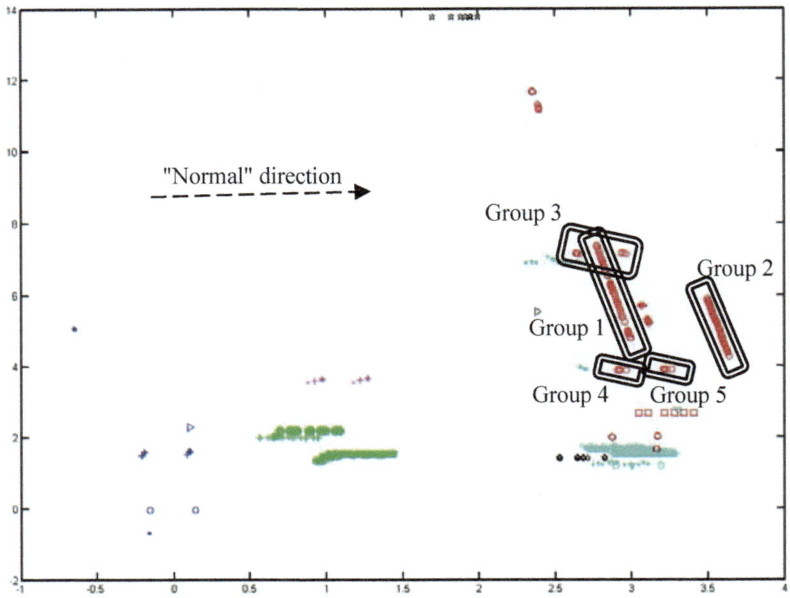

Fig. 5.6 MOVICAB-IDS visualisation of A_3 segment.

Table 5.7 Parameter values of Fig. 5.6.

Parameter values of Fig. 5.6	
Parameter	**Value**
Iterations	120,000
Learning rate	0.01
p	1.125
τ	0.001

Once again, network scans (Groups 1 and 2) and SNMP community searches (Groups 3, 4, and 5) are displayed as lines that are non-parallel to normal traffic in Fig. 5.6.

MOVICAB-IDS visualisation of A13 segment is presented in Fig. 5.7. Apart from the attacks detected in previous accumulated segments (network scans in Group 1 and SNMP community searches in Groups 2, 3, and 4), the MIB information transfers (Groups 5 to 8) can be labelled anomalous in this projection. As in the case of segment S4 (see Fig. 5.4), these anomalous transfers are identified by their non-parallel evolution and their high packet concentrations. Although these anomalous situations are placed in a 106 minute-long accumulated segment containing almost 50,000 packets, these anomalous situations do not slip by unnoticed. This outcome shows the intrinsic robustness of the applied neural model (CMLHL), which is able to respond effectively to a complex dataset.

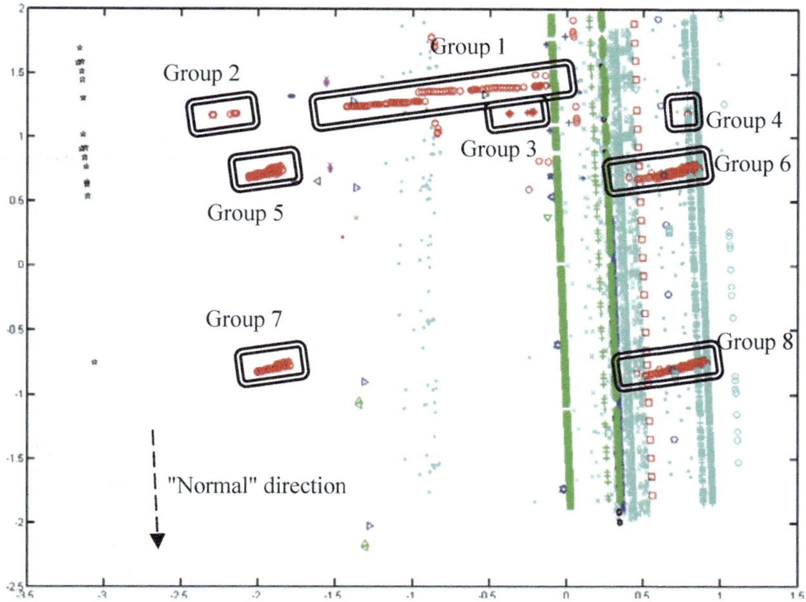

Fig. 5.7 MOVICAB-IDS visualisation of A_{13} segment.

Table 5.8 Parameter values of Fig. 5.7.

Parameter values of Fig. 5.7	
Parameter	**Value**
Iterations	400,000
Learning rate	0.036
p	0.4
τ	0.1

Network scans can be identified within all the accumulated segments in which they are contained (Groups 1 and 2 in Fig. 5.5, Groups 1 and 2 in Fig. 5.6, and Group 1 in Fig. 5.7). Community searches are labelled as Groups 3, 4, and 5 in Fig. 5.6, and Groups 2, 3, and 4 in Fig. 5.7. Additionally, A13 includes two MIB information transfers (Groups 5, 6, 7, and 8 in Fig. 5.7).

As shown in these experiments, MOVICAB-IDS enables the network administrator to identify anomalous situations when packets evolve in non-parallel directions to the "normal" one and when the density of packets is much higher than that of "normal" situations. Empirically, we have noted that an evolution along parallel lines means "normal" traffic data. It has been also shown how traffic evolution is depicted by MOVICAB-IDS, letting the network administrator identify service disruptions.

5.2 DARPA Dataset

This section describes the empirical verification of MOVICAB-IDS involving a port scan attack in the DARPA IDS evaluation dataset [212, 213, 214] made available by the MIT Lincoln Laboratory. Although some works have raised questions about the accuracy and reliability of these datasets [298, 299], the DARPA and the associated KDD Cup 1999 [153, 154] datasets are still the standard corpora for the evaluation of NIDSs.

5.2.1 Dataset Description

The DARPA corpus was assembled in 1998 [1] and 1999 [212, 213] to provide a standard to evaluate both false-alarm rates and detection rates of IDSs, including a variety of known and new attacks buried in a large amount of normal traffic. The corpus was collected from a simulation network used to automatically generate realistic traffic, including attempted attacks. The DARPA corpus provides a widely-used benchmark for ID evaluation on network traffic at the packet level.

In the present study, only TCP (i.e. Transport Control Protocol) traffic was selected from this dataset as most of the attacks (166 out of 174) contained in the 1998 DARPA corpus are based on this protocol. TCP packets contained in a subset of this dataset are characterized by a set of features that has already proved effective in the GICAP-IDS dataset, namely: timestamp, source and destination ports, packet size, and protocol. As such, the TCP network traffic is mapped into a five-dimensional feature space.

5.2.1.1 Port Scans

Among all the SNMP anomalous situations we focused on, only scans are contained in the DARPA dataset. To check the ability of MOVICAB-IDS in facing such attacks through a well known dataset, this section comprises an experimental validation on responding to port scans contained in the DARPA dataset.

Tests in this study involved a subset of the 1998 DARPA dataset. This subset contains 10 minutes of the traffic (3,730 packets) captured on the Monday of the second week, including the portsweep attack generated on that day. In the DARPA documentation page [300], a portsweep attack is defined as a surveillance sweep through many ports to determine which services are supported on a single host. In this sample of portsweep attacks, packets are sent from the host 192.168.1.10 to the 100 first port numbers of the host 172.16.114.50.

5.2.2 Results

In the following figures (Fig. 5.8 and Fig. 5.9), due to the high number of protocols in the DARPA dataset, all the packets except those related to the portsweep (in blue and red) are depicted as black stars.

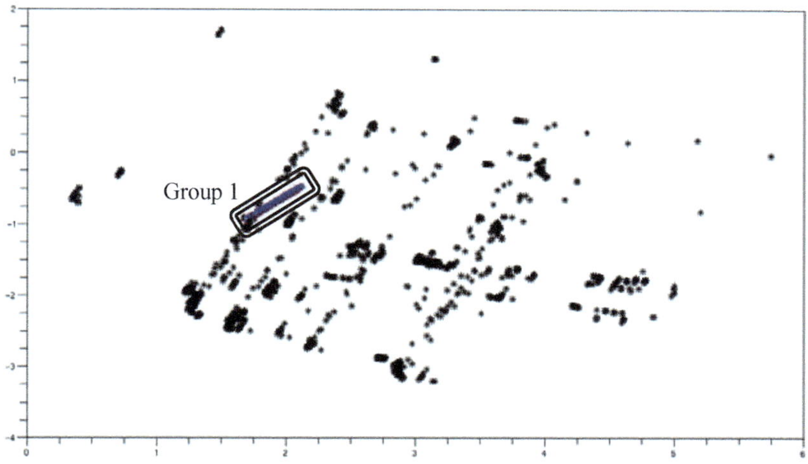

Fig. 5.8 MOVICAB-IDS visualisation of a DARPA sample dataset.

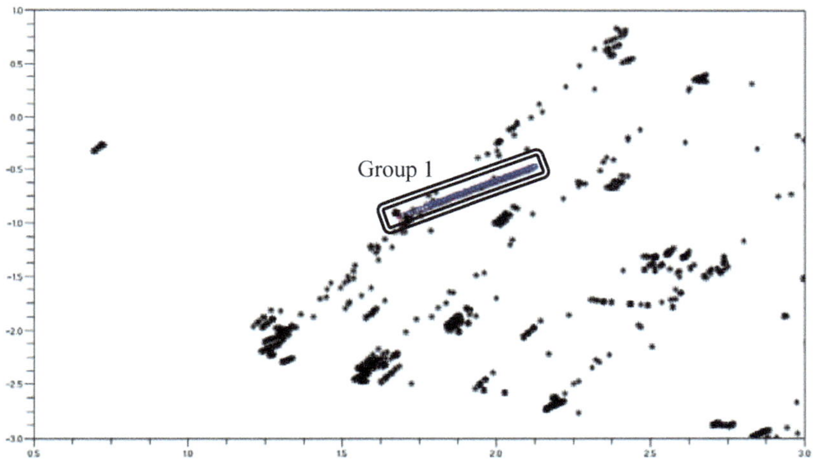

Fig. 5.9 Zoomed portsweep attack.

Fig. 5.8 and Fig. 5.9 show that MOVICAB-IDS manages to identify the anomalous situation contained in the analysed dataset. The portsweep attack (Group 1) is identified due to its non-parallel evolution to the normal traffic. This shows how MOVICAB-IDS is also able to detect anomalous situations in a large dataset following the same previously explained strategy (for the GICAP-IDS dataset). Signs of non-parallel evolution and high packet concentrations are viewed as anomalies.

Chapter 6
Testing and Validation

Some authors claim that one of the main challenges for researchers seeking to implement and to validate a new ID method is to assess and compare its performance with other available approaches [301]. Benchmarking an IDS is a common way of establishing its effectiveness, although this task is neither easy nor clear-cut [16, 302]. It is even more difficult when dealing with visualisation-oriented IDSs as no specific comparative techniques exist, unlike those for other approaches such as the ROC (Receiver Operating Characteristic) analysis [303].

Some interesting works have proposed IDSs testing in real (or as realistic as possible) environments [304] or in experimental settings [82, 93, 282]. These two approaches have proved to have some advantages and disadvantages, as pointed in [305], in which a combination of them is proposed.

The most widely accepted approach to IDS benchmarking is based on a publicly available dataset. Under this approach, the performance of a new IDS can be compared against those of previous IDSs by analysing this dataset. Up to now, only DARPA [1, 199, 200] and the associated KDD Cup 1999 [153, 154] datasets are available for such purpose. Beginning in 1998, DARPA initiated the ID Evaluation program [300]. These evaluations involved generating background traffic interlaced with malicious activity so that IDSs and algorithms could be tested and compared. Some problems concerning the inner structure and features of this dataset have been discovered [299, 306, 307]. It also has some other up-to-date problems such as the outdated attacks it contains and the data rate, that is not comparable with that in current real networks [308]. Despite these limitations, so far, this is still the *de-facto* standard corpus for IDS testing as there are important constraints (privacy and confidentiality mainly) that prevent a new network traffic dataset from being made publicly available. The Lincoln Adaptable Real-time Information Assurance Test-bed (LARIAT) [309] was aimed at easing the ID development and evaluation by extending the DARPA ID Evaluation program. Unfortunately, the LARIAT software is not openly available. MOVICAB-IDS was confronted with some of the anomalous situations contained in the DARPA dataset in section 5.2.

Apart from testing datasets, some methodologies [14, 305, 310], frameworks [78, 93, 94], and tools [145, 228, 281] for assessing IDSs have also been proposed. Additionally, comparative studies on commercial IDSs have been carried out by

Á. Herrero and E. Corchado: Mobile Hybrid Intrusion Detection, SCI 334, pp. 105–121.
springerlink.com

network magazines mainly [304, 311, 312, 313]. However, up to the present, there have been no specific testing proposals for visualisation-oriented IDSs due to the novelty of such approach. Moreover, the projection of network packets onto a 2D/3D space to support visual ID has only been very partially addressed in previous works. Although some approaches to validate IDSs based on categorical data exist [314], little effort has been devoted to ID benchmarks based on numerical traffic data. Despite the fact that increasing effort has been devoted over recent years to IDSs in general, and to IDSs assessment in particular, there is no comprehensive and scientifically rigorous methodology to test the effectiveness of these systems [315]. Thus, it remains an open issue and a significant challenge [301].

In view of the above-mentioned issues, a fair comparison between MOVICAB-IDS and other similar visualisation methods for ID is difficult to achieve. In consequence, to measure the performance of MOVICAB-IDS, a two-fold analysis has been designed and carried out. This analysis consists of:

- **Mutation-based technique** (section 4.6): a novel testing technique that specializes in IDSs relying on numerical packet features. It is based on measuring the evaluated IDS results when confronting unknown anomalous situations, as the identification of 0-day (previously unseen) attacks is possibly the most challenging and important problem in ID [27].
- **Comparison** (section 4.7): various projection/mapping models are compared when facing packet visualisation for ID.

In this chapter, the analysis is described and the most relevant results are shown. MOVICAB-IDS is then qualitatively measured against a predictable baseline and its subjacent neural model is comparatively measured against other projection/mapping techniques [302]. The main limitations of the proposed analysis are the absence of quantifiable outcomes and the dependence on previously captured background and attack traffic.

6.1 Mutation Testing Technique

The mutation testing technique that has been developed is based on a measured evaluation of the IDS results after having confronted unknown anomalous situations, as the identification of 0-day (previously unseen) attacks is a key issue in ID. Some ID strategies (especially those based on attack signatures or patterns) can not properly deal with such attacks. The goal is to test an IDS in situations that could occur with a hacker that are as real as possible.

Inspired by previous attempts [316, 317], this testing model consists of generating "synthetic" attacks by mutating "real" attacks on a background of "normal" traffic. In other words, attacks are injected into a stream of real background activity. This is a very effective approach for determining the hit rate of an IDS given a particular level of background activity [315]. The main advantage of such an approach is that as the background activity is real, it contains all of the "usual" anomalies and subtleties.

In general, a mutation can be defined as a random change. In keeping with this idea, the testing technique changes different features of the packets belonging to a known attack to generate previously unseen attacks. As previously explained in Chapter 4, the numerical information to be mutated is extracted from the packet headers. As a result, the IDS will be tested in "semi-real" situations that may be generated by a hacker, which would not otherwise be available for training the IDS.

The modifications created by this testing technique lead to real situations by involving changes in aspects such as: attack length (amount of time that an attack lasts), packet density (number of packets per time unit), attack density (number of attacks per time unit), and time intervals between attacks. The mutations can also concern both source and destination ports, varying between the three different ranges of TCP/UDP port numbers: well known (from 0 to 1023), registered (from 1024 to 49151), and dynamic/private (from 49152 to 65535). Some of the possible mutations may be senseless, such as a scan of less than three hosts in the case of a network scan, which was also taken into account. Additionally, certain restrictions are imposed on the mutations in order to generate new attacks that are as realistic as possible: time values must range from 1 to the length of the original segment while port numbers must range from 0 to 65535.

Time is a fascinating issue of great importance when considering intrusions since the longer an attack is, the more likely its detection. There are therefore two main strategies that will allow an intrusion to slip by unnoticed:

- Drastically reduce the time used to perform an attack, by increasing the temporal concentration of packets.
- Spread the packets out over time, which is to say, reduce the number of packets sent per time unit.

These strategies have been taken into account when applying the proposed mutation technique.

6.1.1 Mutating a Sample Dataset

The novel testing technique was verified by applying it to a sample dataset generated on the same network on which the experimental study of MOVICAB-IDS had been performed. See Chapter 5 for further details.

This sample dataset contains the following anomalous situations:

- Three network scans to several hosts aimed at port numbers 161, 162, and 3750.
- A time difference between the first and the last packet included in each scan (i.e. scan length) of 17,866 ms for the scan aimed at port number 161, 22,773 ms for the scan aimed at port number 162, and 17,755 ms for the scan aimed at port number 3750.
- An MIB information transfer.

For ease of comparison, the MOVICAB-IDS visualisation of the original dataset is shown in Fig. 6.1.

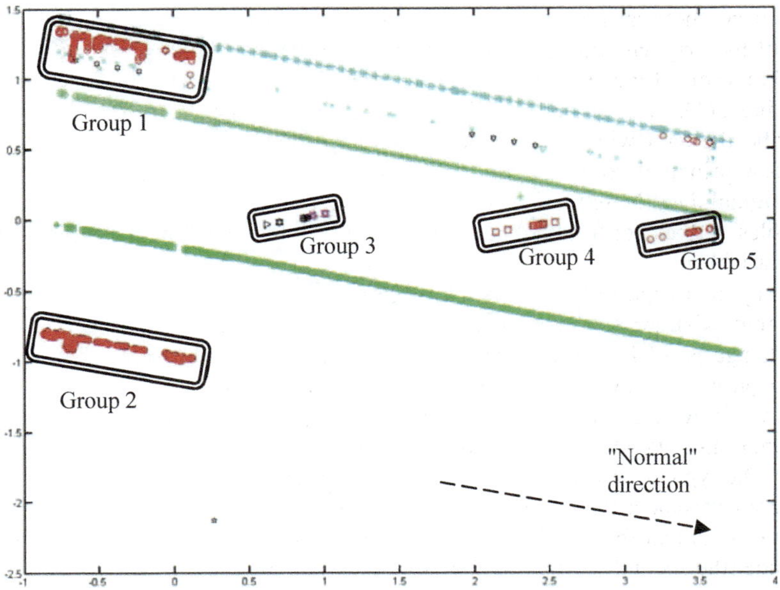

Fig. 6.1 MOVICAB-IDS visualisation of the original dataset to be mutated.

As may be observed in Fig. 6.1 and as previously explained in 5, MOVICAB-IDS identifies anomalous situations due to the fact that these situations do not tend to resemble parallel and smooth directions (as normal situations do) or due to their high packet concentrations. Accordingly, the network scans (Groups 3, 4, and 5 in Fig. 6.1) and the MIB information transfer (Groups 1 and 2 in Fig. 6.1) in the sample dataset can be labelled anomalous.

6.1.1.1 Dataset Description

Several tests were initially designed to verify the performance of MOVICAB-IDS through the mutation testing technique. Each one of these tests is associated with a dataset obtained by mutating the original one. Mutations relating to network scans have been performed and changes were therefore made to the scan packets to take the following issues into account:

- Number of scans in the dataset: i.e. number of scanned ports.
- Destination port numbers targeted by the scans.
- Length of the scans.
- Number of packets (density) in the scans. That is, number of scanned hosts.

By considering these issues, the collection of testing datasets covers most of the different scan-related situations which the monitored network might have to confront. Testing datasets can then be roughly described as:

- **Mutated dataset 1:** one scan aimed at port 3750.
- **Mutated dataset 2:** two scans aimed at ports 161 and 162.

- **Mutated dataset 3:** one scan aimed at port 1734.
- **Mutated dataset 4:** two scans aimed at ports 4427 and 4439.
- **Mutated dataset 5:** three time-expanded scans aimed at ports 161, 162, and 3750.
- **Mutated dataset 6:** three time-contracted scans aimed at ports 161, 162, and 3750.
- **Mutated dataset 7:** one time-expanded scan aimed at port 3750.
- **Mutated dataset 8:** two 5-packet scans aimed at ports 4427 and 4439.
- **Mutated dataset 9:** two 30-packet scans aimed at ports 1434 and 65788.

The first issue to consider is the number of scans in the attack. Datasets containing one scan (mutated datasets 1, 3, and 7), two scans (mutated datasets 2, 4, 8, and 9) or three scans (mutated datasets 5 and 6) were generated. Each one of the scans in a dataset is aimed at a different port number. The underlying idea is that hackers can check the vulnerability of as many services/protocols as they want. The number of scans (ranging from 1 to 65536) can be modified from one attack to another.

A scan that attempts to check whether a protocol/service is running can be aimed at any port number (from 0 to 65535). The mutated datasets that were introduced above contained scans aimed at port numbers such as 161 and 162 (well known ports assigned to SNMP), 1434 (registered port assigned to Microsoft-SQL-Monitor, the target of the W32.SQLExp.Worm), 3750 (registered port assigned to CBOS/IP ncapsalation), 4427 and 4439 (registered ports, as yet unassigned), and 65788 (dynamic or private port). Then, the different port ranges were covered by these samples of attacks.

Mutated datasets 5, 6, and 7 were used in order to check the proposed IDS in relation to the time-related strategies. Mutated dataset 5 was obtained by spreading the packets contained in the three different scans (aimed at port numbers 161, 162, and 3750) over the captured data. In this dataset, there was a time difference of 247,360 ms between the first scan packet (aimed at port 161) and the last scan packet (aimed at port 3750), whereas in the original dataset the scans lasted 164,907 ms. The total length of the mutated dataset was 262,198 ms. In the case of mutated dataset 7, the same mutation was performed but only for those packets that were associated with the scan aimed at port 3750. In contrast, the strategy of reducing the attack time was followed to obtain mutated dataset 6. In this case, the time difference between the first and the last packet of the scans was about 109,938 ms.

Finally, the number of packets contained in each scan was considered. In the case of a network scan, each packet means a different host is included in the scan. Mutated datasets 8 and 9 were created to address this issue. Mutated dataset 8 contains low-density scans given that were reduced to only five packets. It was decided that a scan probing less than five hosts could not be considered a proper network scan. This is a fuzzy lower limit as it could also be set as four or six packets. On the other hand, mutated dataset 9 contained medium-density scans. In this case, each one of them was extended to 28 packets.

Apart from identifying the mutated scans, the detection of the MIB information transfer contained in all these datasets also represented a test for the performance of MOVICAB-IDS. The experimental results obtained for these mutated datasets are shown in the following section.

6.1.1.2 Results

The application of MOVICAB-IDS to the different scenarios described in section 6.1.1.1 led to the results shown in this section. For the sake of brevity, only visualisations of the most representative cases are presented.

As previously stated in Chapter 5, scans are labelled anomalous whenever their packets tend not to resemble parallel and smooth directions (normal situations). Additionally, other anomalous situations, such as MIB information transfers, are also identified by CMLHL as anomalous on the basis of traffic density.

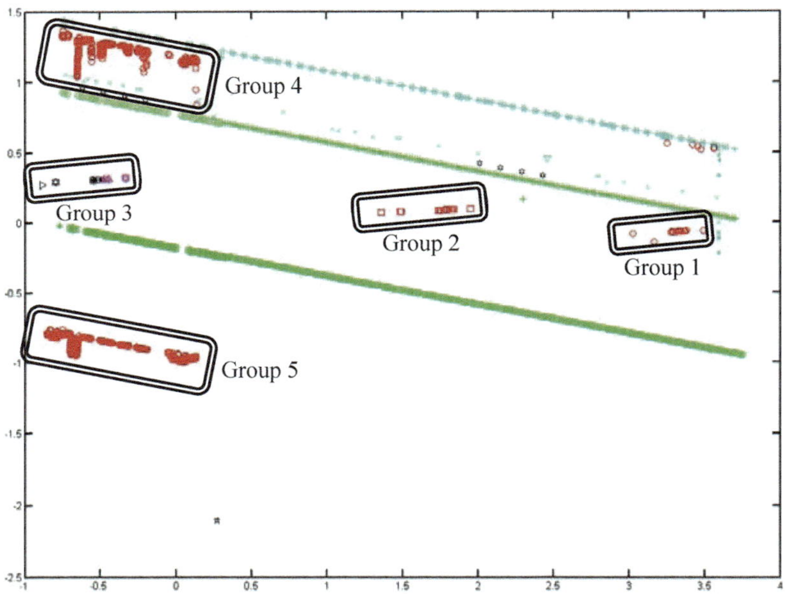

Fig. 6.2 MOVICAB-IDS visualisation of mutated dataset 5.

The visualisation of mutated dataset 5 is shown in Fig. 6.2. The three time-expanded scans aimed at ports 161, 162, and 3750 can be easily identified (Groups 1, 2, and 3 respectively). Additionally, the MIB transfer can also be more clearly identified (Groups 4 and 5).

The three time-contracted scans (Groups 1, 2, and 3) contained in mutated dataset 6 can be easily identified in Fig. 6.3. On the contrary, considerable experience is required to identify the time-expanded scan (Group 1) in the case of the visualisation of mutated dataset 7 (Fig. 6.4). As previously mentioned, this would

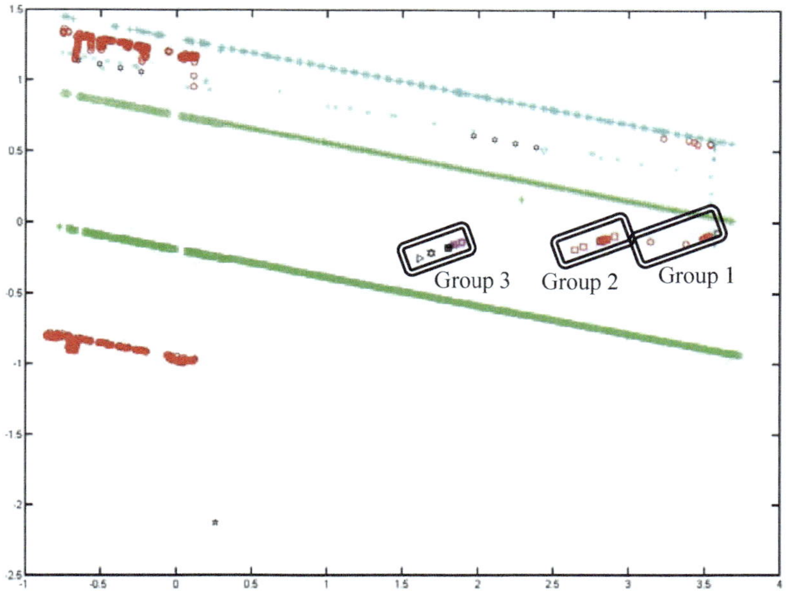

Fig. 6.3 MOVICAB-IDS visualisation of mutated dataset 6.

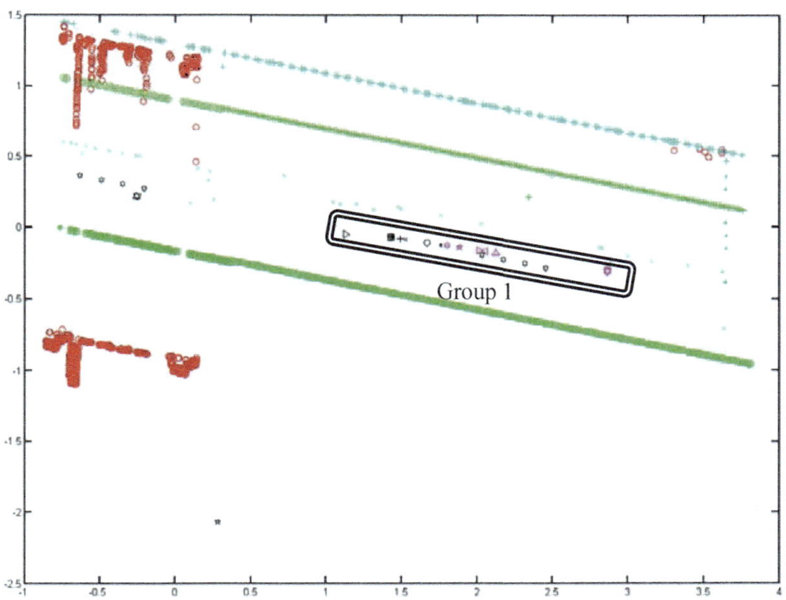

Fig. 6.4 MOVICAB-IDS visualisation of mutated dataset 7.

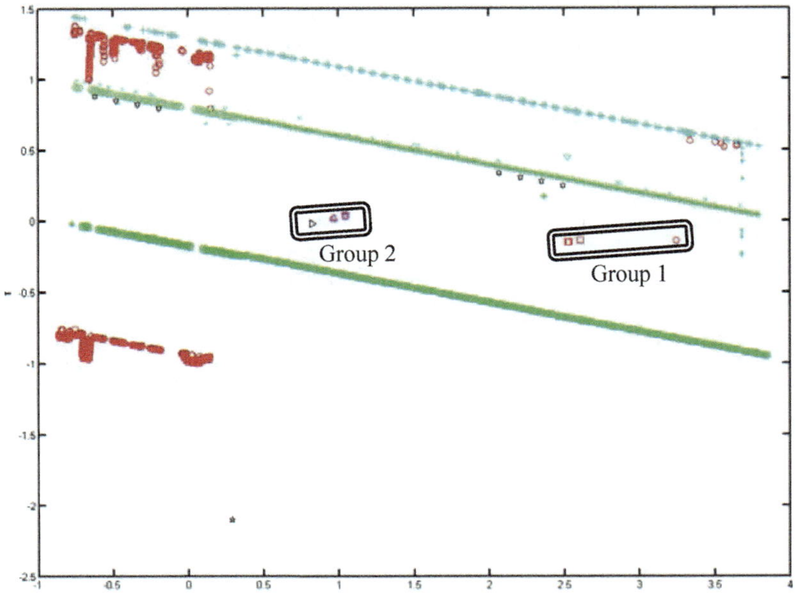

Fig. 6.5 MOVICAB-IDS visualisation of mutated dataset 8.

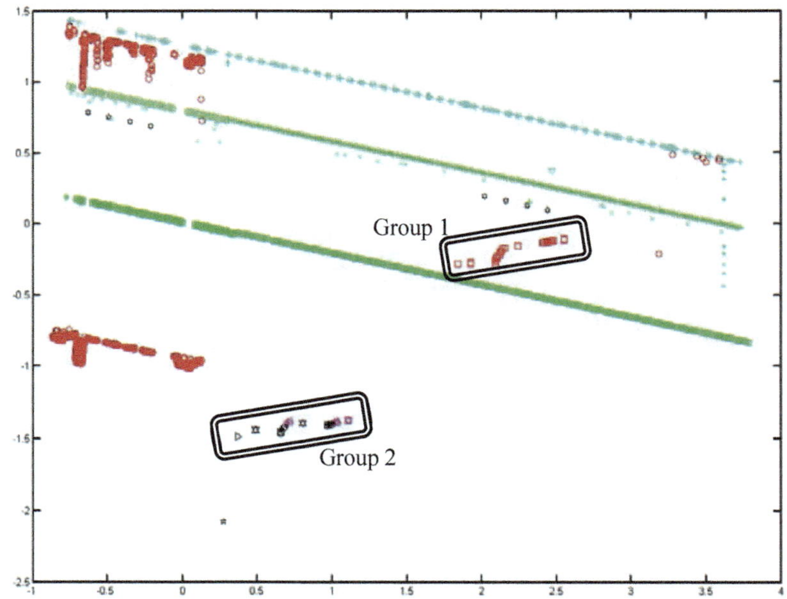

Fig. 6.6 MOVICAB-IDS visualisation of mutated dataset 9.

correspond to a hacker trying to slip by unnoticed by spreading the attack packets out over time. It is worth emphasizing that the other anomalous situation (the MIB information transfer) was identified in mutated dataset 7 with greater clarity than in any of the other visualisations of the mutated datasets.

When scans contain only five packets (Groups 1 and 2 in Fig. 6.5 - mutated dataset 8), an expert is once again required to identify the anomalous scan situations. On the other hand, MOVICAB-IDS clearly detects high-density scans as the ones contained in mutated dataset 9 (Groups 1 and 2 in Fig. 6.6).

Through the above figures (from Fig. 6.2 to Fig. 6.6), it may be seen how MOVICAB-IDS is able to identify the different mutated anomalous situations, even though some are identified with greater clarity than others.

These initial tests demonstrate the ability of MOVICAB-IDS to identify the anomalous situations it confronted. Identification of the mutated scans can, in broad terms, be explained by the generalization capability of the neural model in the proposed IDS. In other words, through the use of CMLHL, MOVICAB-IDS is capable of identifying not only the real anomalous situations contained in the original dataset (known) but also the mutated (unknown) ones which may be real.

6.1.2 Mutating Segments

After mutating a sample dataset, the proposed testing technique was applied to more complex datasets: accumulated segments previously analysed by MOVICAB-IDS (see Chapter 5). Each one of the mutations generates a new segment or dataset. For the sake of simplicity, only the visualisations of two mutated datasets (A2' and A2") are shown in this section. These mutated datasets were generated from the A2 accumulated segment (see Chapter 5).

Fig. 6.7 shows the MOVICAB-IDS visualisation of a mutated version of the A2 accumulated segment (A2'). This mutation implies changes in the destination port numbers of the network scans contained in the original dataset. The network scan, originally (in A2) aimed at port number 1434, is now (in A2') aimed at port number 23745. In the second network scan, the original destination port number 65788 was replaced by 45232 (in A2'). Additionally, the packet density of these scans was reduced; each of the original scans consisted of 60 packets which were reduced to 40 packets in A2'.

Fig. 6.8 shows the MOVICAB-IDS visualisation of a different mutated version (A2") of the A2 accumulated segment. In this case, only one network scan is kept (scan aimed at port number 1434 in A2 is removed) and the packet density of this scan is decreased by enlarging the duration of the attack. In the original dataset (A2), the network scan lasted 61,858 ms, while in this mutated dataset (A2"), it lasted 247,432 ms. This mutation could correspond to the strategy of an attacker trying to slip by unnoticed by reducing the attack time.

To check the ability of MOVICAB-IDS in identifying 0-day attacks, these mutated segments (A2' and A2") were projected through the weights previously calculated for the A2 segment. This simulates a 0-day (previously unseen) attack as the neural model was trained without these data. The projections then revealed whether MOVICAB-IDS had been able to identify the previously unseen attacks.

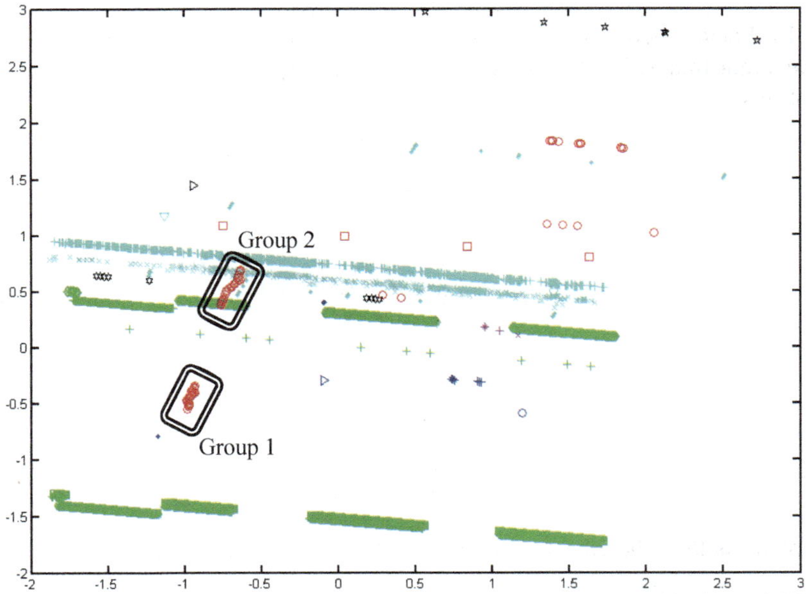

Fig. 6.7 MOVICAB-IDS visualisation of A_2' mutated segment.

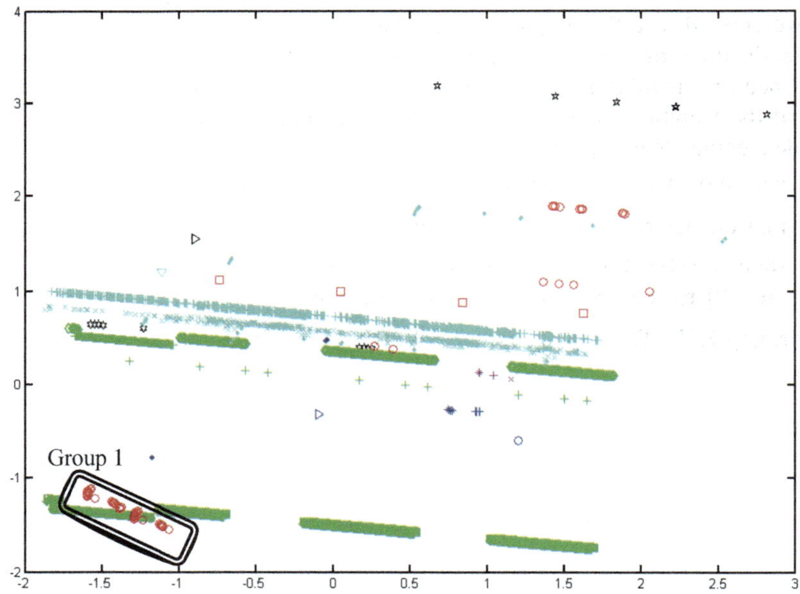

Fig. 6.8 MOVICAB-IDS visualisation of A_2'' mutated segment.

The network scans contained in both A2' (Groups 1 and 2 in Fig. 6.7) and A2'' (Group 1 in Fig. 6.8) segments are labelled anomalous due to their non-parallel evolution to normal traffic. It is less easy to identify the network scan in the case of the A2'' segment due to its lower packet density, which was expected, as higher packet densities signal anomalies.

The projections of the mutated datasets can be compared with the projection of the original A2 segment (Fig. 5.5). By doing this, we can say that due to the generalization capability of the neural model underlying MOVICAB-IDS, it is able to identify previously unseen network scans. Thus, we can conclude that the proposed mutation testing technique positively evaluates MOVICAB-IDS.

6.2 Comparison with Other Projection Models

After applying mutant testing, it was decided to compare the outcome of the MOVICAB-IDS underlying model, CMLHL, with those of other well-known statistical and connectionist models. Both linear and nonlinear models such as PCA, CCA, and SOM (described in section 2.5) have been applied to some of the previously analysed datasets (see Chapter 5). Several experiments were conducted to apply these models to the segments in the GICAP-IDS dataset. The performance of CMLHL in visualising other network traffic datasets has also been compared to other projection models (such as Linear Discriminant Analysis and Nonlinear PCA) in previously published papers [283, 318].

For the sake of simplicity, only projections concerning two segments (A2 and A3) previously analysed by CMLHL (see section 5.1) are provided in this section. The computation of mutual distances when applying CCA, which is highly resource demanding, prevents larger datasets from being used. For these segments, only the best results (from the visualisation point of view) obtained after tuning the parameters of the models are shown. See Fig. 5.5 (A2 segment) and Fig. 5.6 (A3 segment) for the CMLHL projection of these datasets.

In the case of CCA, some parameters such as alpha, lambda, number of epochs, and distance criterion were tuned. In the case of the SOM, the following options and parameters were tuned: grid size, lattice, batch/online training, initialization, and neighbourhood function. The experiments concerning the SOM were performed by means of the SOM Toolbox for Matlab [319].

6.2.1 Principal Component Analysis

The statistical technique known as PCA (described in section 2.5.6) was applied to the A2 segment (Fig. 6.9). This technique, already used in the field of IDSs [249], failed to detect the anomalous situations (network scans), although the two principal components (depicted in Fig. 6.9) amount to 99.99% of the data's variance. None of the network scans (Groups 1 and 2 in Fig. 6.9) contained in A2 were identified as anomalous traffic because in this projection all the packets evolve in parallel lines (which is associated to normal traffic in this work). In the ID jargon, these will be false negatives.

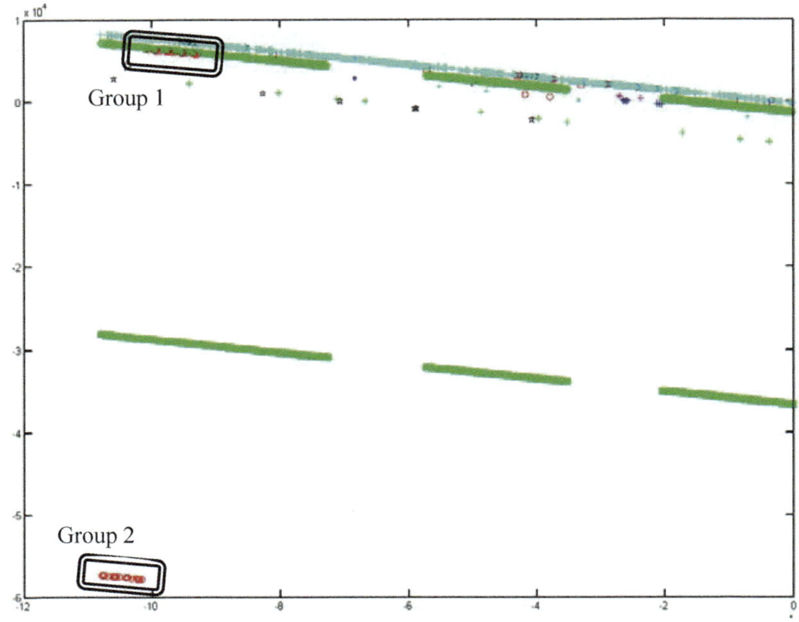

Fig. 6.9 PCA projection of A_2 segment.

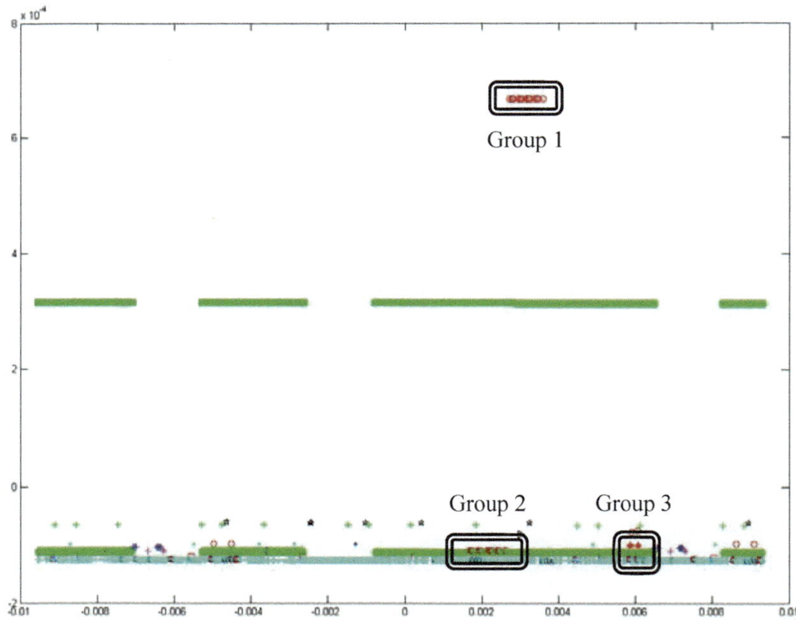

Fig. 6.10 PCA projection of A_3 segment.

PCA was also applied to the A3 segment (Fig. 6.10). This projection, amounting to 99.9% of the data's variance, failed to detect the anomalous situations (network scans and SNMP community search). Neither of the network scans (Groups 1 and 2) and the SNMP community searches (Group 3) contained in A3 were identified as anomalous traffic because in this projection their packets evolve in parallel lines.

6.2.2 Curvilinear Component Analysis

Fig. 6.11 shows the projection of A2 obtained by CCA (described in section 2.5.14) using standard Euclidean distance. The other final selected parameter values were: lambda = 168,850 (default value), alpha = 0.3, and 7 epochs. The anomalies could be differentiated from normal traffic on the basis of parallel evolution as in the case of CMLHL. Network scans (Groups 1 and 2 in Fig. 6.11) are depicted in a non-parallel way to normal traffic. The main difference between the CMLHL projection and the one obtained by CCA is that the non-parallel visualisation provided by CMLHL is much clearer. In the case of CCA, some of the groups containing normal traffic are depicted in a similar way to those containing the network scans and could therefore constitute false positives. Identification of the anomalous situations is therefore less clear than in the CMLHL projection.

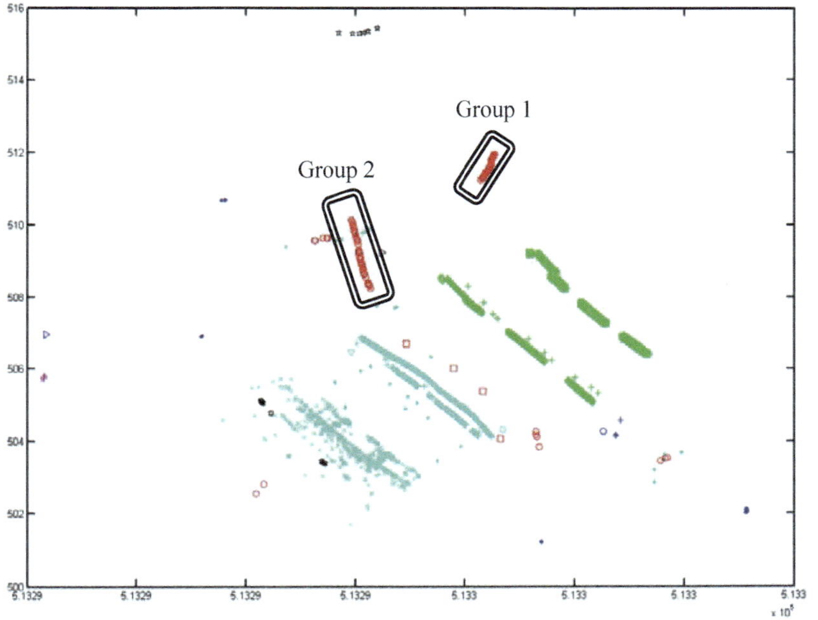

Fig. 6.11 CCA projection of A_2 segment.

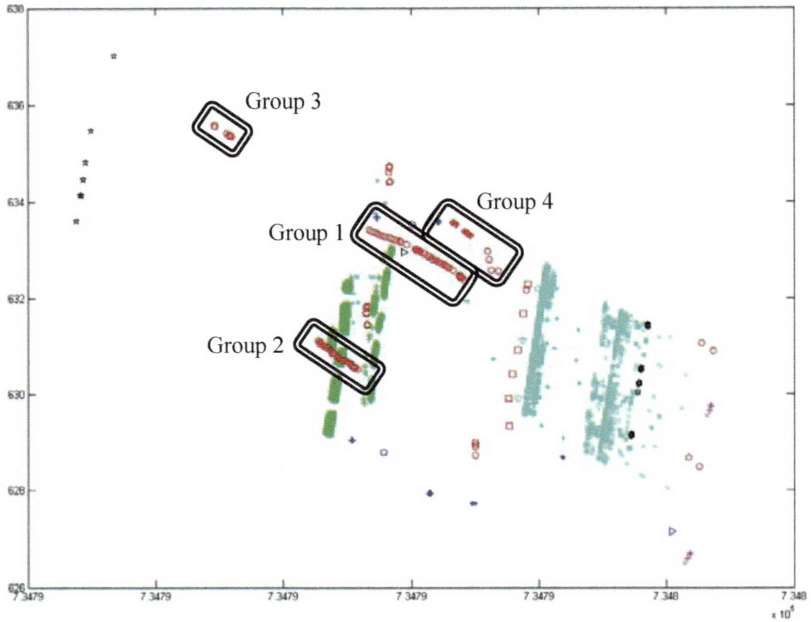

Fig. 6.12 CCA projection of A_3 segment.

Fig. 6.12 shows the projection of A3 obtained by CCA using standard Euclidean distance. The other final selected parameter values were: lambda = 100,000, alpha = 0.6, and 9 epochs. The network scans (Groups 1 and 2 in Fig. 6.12) together with the SNMP community searches (Groups 3 and 4 in Fig. 6.12) are depicted as lines that are non parallel to normal traffic. Although both CMLHL and CCA provide quite interesting projections of A3 segment, the main difference between these projections is that CMLHL minimizes the overlapping between the different groups, yielding clearer and sparser projections.

6.2.3 Self-Organizing Map

The SOM mapping (described in section 2.5.13) of the A2 segment is depicted in Fig. 6.13. The quality measures associated with this SOM mapping are: quantization error = 0.016 and topographic error = 0.073.

As the SOM is unable to process a linear growing variable properly, such as the timestamp, this information has been removed from the dataset. For visualisation purposes, all the packets in the analysed segment were labelled according to the following labels: C1 for normal traffic, C2 for the network scan aimed at port number 1434, and C3 for the network scan aimed at port number 65788.

The used parameter values for this mapping were: linear initialization, batch training, hexagonal lattice, and Gaussian neighbourhood function. The grid size (13x29) was determined by means of a heuristic formula.

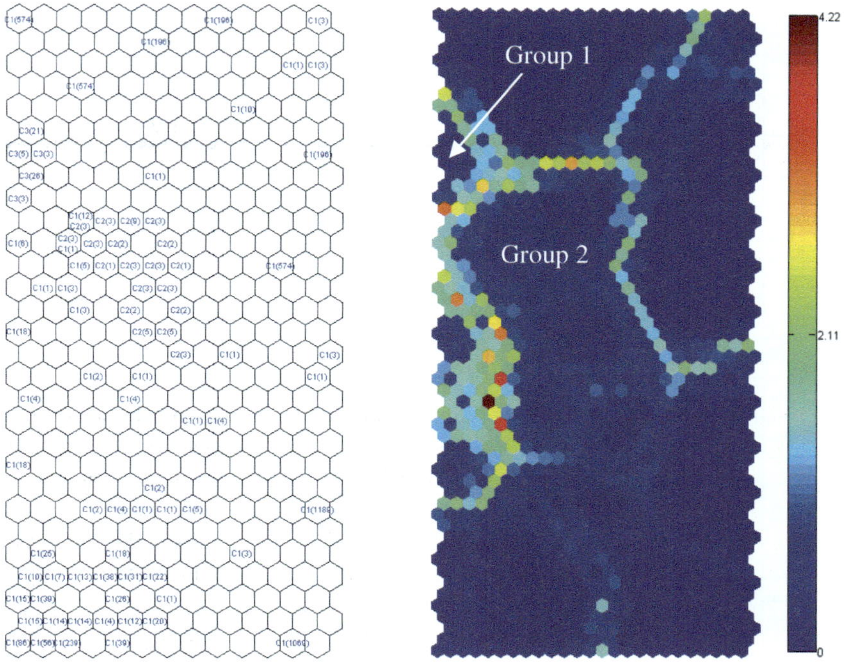

a. Labelled map.

b. Associated U-matrix.

Fig. 6.13 SOM visualisation of A_2 segment.

Fig. 6.13.a depicts the labelled generated map which shows the labels of the instances (packets) identified by each neuron. The number of instances assigned to each neuron appears in parentheses. The associated U-matrix, which depicts the distances between the neurons in the lattice, is shown in Fig. 6.13.b.

It can be clearly seen that the SOM is able to cluster the presented segment. The group of neurons near the upper-left corner of the lattice (Group 1 in Fig. 6.13.b) gathers all the packets labelled as C3, that is, all the packets belonging to the network scan aimed at port number 65788. No other class of traffic is identified by the neighbouring neurons. Another important group can be identified in the middle of the lattice (Group 2 in Fig. 6.13.b). This group gathers all the packets associated with the other anomalous situation (labelled as C2 in Fig. 6.13.a). Additionally, some packets associated with normal traffic are also identified by these neurons. By using the SOM mapping, normal traffic could be identified as belonging to an anomalous situation (false positive) and the remaining traffic is identified by other different neurons.

Finally, the SOM mapping of the A3 segment is depicted in Fig. 6.14. The quality measures associated with these SOM results are: quantization error = 0.122 and topographic error = 0.11.

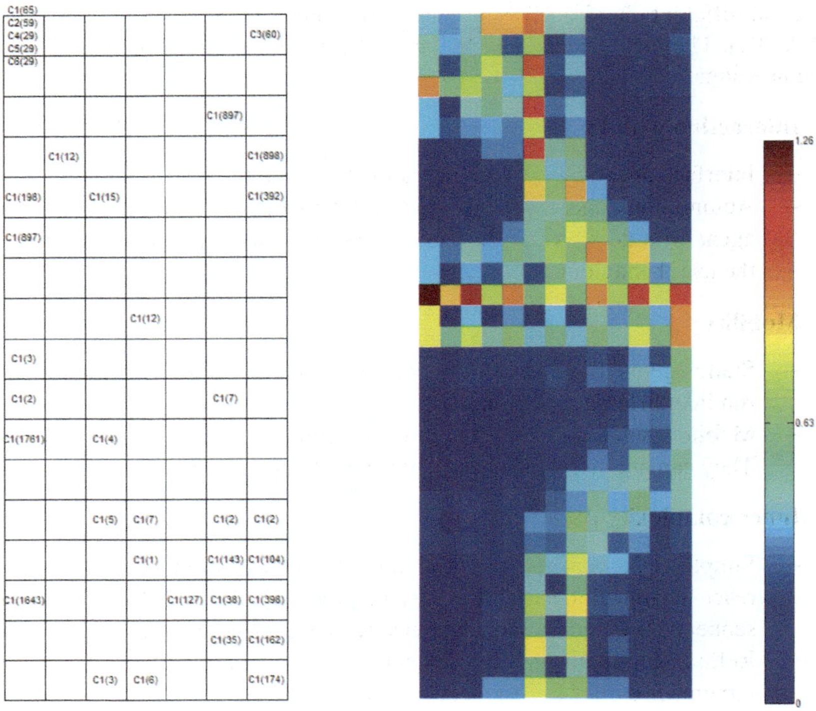

a. Labelled map. b. Associated U-matrix.

Fig. 6.14 SOM visualisation of A_3 segment.

For visualisation purposes, all the packets in the analysed segment were labelled according to the following classes: C1 for normal traffic, C2 for the network scan aimed at port number 1434, C3 for the network scan aimed at the port number 65788, C4 for the community search aimed at port number 161, C5 for the community search aimed at port number 1161, and C6 for the community search aimed at port number 2161.

The used parameter values for such mapping were: random initialization, sequential training, rectangular lattice, and bubble neighbourhood function. The grid size (7x17) was determined by means of a heuristic formula.

Fig. 6.14.a depicts the labelled map that was generated in which the classes identified by each neuron are shown. The number of instances (packets) belonging to each one of the classes identified by the neurons are shown in parentheses. Fig. 6.14.b shows the associated U-matrix.

It can be seen that the neuron in the upper-right corner of the rectangular lattice (Fig. 6.14.a) gathers all the packets labelled as C3, that is, all the packets belonging to the network scan aimed at port number 65788. No other class of traffic is identified by the neighbouring neurons. The neuron in the upper-left corner of the lattice gathers all the packets associated with the other anomalous situations

(C2, C4, C5, and C6). Additionally, 65 packets associated with normal traffic were also identified by this neuron, which once again proved to be false positives. The rest of the normal traffic is identified by other different neurons.

It may be briefly concluded that some projection models such as PCA, EPP, MLHL, CMLHL, and CCA have one important advantage over other unsupervised neural models (such as the SOM) within the framework of ID: they use time as a key variable when analysing the evolution of the packets in the traffic dataset. This allows the depiction of packets in an intuitive way, which is especially valuable in the field of visual inspection of network traffic for ID.

Chapter 7
Discussion and Conclusions

This chapter presents a discussion of the results obtained by applying the proposed IDS to different datasets and the results of the testing study, which are respectively reported in Chapter 5 and Chapter 6. It reports the principal conclusions of the research work in this book and sets out future lines of research.

7.1 Discussion

Having examined the projections obtained in the experiments shown in Chapter 5, it can be said that MOVICAB-IDS visualisation is able to depict normal traffic in a compact and intuitive way, as can be seen in Fig. 5.1. This facilitates the task of network monitoring, as security personnel are given a picture of the network status at a glance. Additionally, this visualisation allows the identification of traffic disruption associated with certain protocols which can be the consequence of attacks (such as Denial of Service) or malfunctioning.

From the other experiments presented in Chapter 4 (from Fig. 5.2 to Fig. 5.7), it can be seen that network scans are depicted as non-parallel lines to the direction in which normal traffic evolves. This is the "visual signature" of such attacks, which anybody without in-depth knowledge of the visualisation technique can understand as a signal that something anomalous is taking place in the network. SNMP community searches are depicted in a similar way to network scans. It is therefore not easy to discriminate between either of the two types of attacks, to establish exactly which is taking place at any one time. To do so, security personnel have to check further information, such as the headers of the packets identified as anomalous or the outputs of some other security tools. Finally, MIB information transfers may be easily identified by security personnel as they have a unique "visual signature" which reflects the high packet density and non-parallel evolution.

It may be noted from the projections of accumulated segments that the more packets a dataset contains, the more difficult it is to distinguish anomalous situations. This is mainly due to the overplotting weakness of scatter plots (described in section 2.4), which is overcome by splitting data into simple segments. As can be seen in section 5.1, anomalous situations are more clearly identified in simple segments, which eases the ID task. On the other hand, accumulated segments

Á. Herrero and E. Corchado: Mobile Hybrid Intrusion Detection, SCI 334, pp. 123–128.
springerlink.com

provide a more comprehensive view of network traffic, which facilitates the network monitoring task. When applied to high-volume networks, MOVICAB-IDS can deal with massive datasets (such as the DARPA dataset analysed in section 5.2) thanks to this traffic segmentation.

Despite the fact that the MOVICAB-IDS visualisation may not provide as much information as other tools do, it does provide a highly intuitive visualisation of traffic evolution. Wide experience or extensive training on the tool is not required to detect and identify anomalous situations. By using MOVICAB-IDS, inexperienced security personnel can identify anomalous situations merely by taking a quick look at the CMLHL projections. One of the key issues of such an advantage is the preservation of the temporal context of packets, which helps to provide an idea about the network status.

The results of the testing study prove the generalisation capability of MOVICAB-IDS, which shows the "visual patterns" of novel attacks even if the neural model has not faced such attacks earlier. It should be noted that attackers employ very different strategies to go unnoticed. Spreading the attack packets over time (Fig. 6.4) or reducing the amount of packets in an attack (Fig. 6.5) will cause MOVICAB-IDS to visualise these attacks less clearly.

By comparing CMLHL to different unsupervised neural models (section 6.2), it can be concluded that PCA is not able to identify any of the anomalous situations under study. On the contrary, CCA can identify these situations but less clearly than CMLHL. Furthermore, CCA is much more resource demanding than CMLHL, as the former technique needs to compute the pairwise distance matrix for the whole dataset. The SOM is able to distinguish between anomalies and normal traffic but with a low level of accuracy. As a consequence, CMLHL was chosen as the projection model for MOVICAB-IDS.

In addition to purely SNMP anomalous situations (SNMP community searches and MIB transfers), MOVICAB-IDS also helps in detecting probing attacks (network/port scans aimed at any port numbers and hosts) as is evident from the projections that have been shown. Additionally, MOVICAB-IDS traffic snapshots also provide visual insight into network functioning: status of protocols, protocol interactions, traffic volume and diversity, etc. All these features can be roughly identified by visual exploration.

Some existing IDSs need a "clean" (free of attacks) training dataset. This is not the case of MOVICAB-IDS, which can be trained with a dataset containing known and/or new attacks, due to the generalisation capability of the underlying neural model. This is the case of the mutated segments shown in section 6.1. The identification of the mutated scans can, in broad terms, be explained by the generalization capability of the neural model in the proposed IDS.

Due to information privacy and legal concerns, organisations often "sanitise" data before sharing it. By doing so, sensitive information such as IP addresses, packet payload, and so on, is removed. As has been shown, MOVICAB-IDS is able to detect some intrusions without using such sensitive information. The model is able to identify anomalous situations without using either the physical address (MAC) or the IP address. On the other hand, using this reduced set of packet information does not prevent it from detecting some common and dangerous internal attacks from within the network: those related to SNMP.

7.2 Conclusions

The identification of patterns that exist across dimensional boundaries in high dimensional datasets is a fascinating task. Such patterns may become visible if changes are made to the spatial coordinates. However, an *a priori* decision as to which parameters will reveal most patterns requires prior knowledge of unknown patterns.

Projection methods project high-dimensional data points onto a lower dimensional space in order to identify "interesting" directions in terms of any specific index or projection. Having identified the most interesting projections, the data are then projected onto a lower dimensional subspace plotted in two or three dimensions, which makes it possible to examine the structure with the naked eye. Projection methods can be seen as smart compression tools that map raw, high-dimensional data onto two or three dimensional spaces for subsequent graphical display. In that way, the structure identified through a multivariable dataset may be visually analysed with greater ease.

An IDS can be useless if nobody is looking at its output [20]. To that end, projection techniques can map high-dimensional data onto a low-dimensional space adaptively, for the user-friendly visualisation of monitored network traffic. In keeping with this idea, MOVICAB-IDS goes one step further than previous visualisation-based IDSs, in that it combines all the features extracted from the packet headers to depict each simple packet by means of a neural projection model. Thus, it provides the security personnel with a general snapshot of network traffic, protocol interactions, and traffic volume in order for them to identify anomalous situations. Scatter plots have been chosen due to their relative simplicity in comparison to other multidimensional visualisation techniques, familiarity among users, and their high visual clarity [51].

Generally speaking, small and medium-sized enterprises do not employ a network administrator to identify possible threats in the corporate network. On the other hand, network administrators from large companies are usually overwhelmed by information coming from IDS, especially due to the high false-alarm rate. The need for visual support for ID was identified in [11]: "*visualisation tools need to be designed so that anomalies can be easily flagged for later analysis by more experienced analysts*". Visualisation tools can therefore contribute to the security tasks in the following way:

- Visualisation tools may be understood intuitively (even for inexperienced staff) and require reduced configuration time.
- Providing an intuitive visualisation of network traffic lets inexperienced security staff know how normal traffic behaves, which is a key issue in ID [320]. The monitoring task (usually the least dependent on the expertise of the analysts) can be then assigned to less experienced security staff. Thus, workloads will be reduced for senior staff while expertise will be fostered in less-experienced employees.

- As stated in [174], "*visualizations that depict patterns in massive amounts of data, and methods for interacting with those visualizations can help analysts prepare for unforeseen events*". Hence, such tools can also be used in security training.
- They can work in unison with some other tools for ID in a complementary way.

As with other machine learning paradigms, the interesting facet of ANNs learning is not just that the input patterns may be learned/classified/identified precisely but that this learning can be generalised. Whereas learning takes place on a set of training patterns, an important property of the learning process is that the network can generalise its results on a set of test patterns that were not previously learnt. The identification of unknown patterns fits the 0-day attacks [37] detection, which still is very much one of the open issues in the field of ID.

Anomaly-based IDSs model "normal" traffic patterns and generate alerts when "abnormal" events are detected. These techniques do not embed sets of rules and can support time-zero detection of novel attack strategies. On the other hand, the anomaly-based approach requires consistent modelling of normal traffic. Accuracy in detection proves to be the major limitation of anomaly-based IDSs, which can exhibit a relatively high rate of false positives [37]. The typical approach to circumvent that issue is to drive the IDSs development empirically, and connectionist models can be profitably used for that purpose. Most supervised methods [314, 321, 322, 323] tackle intrusion detection as a binary classification problem (i.e., normal/anomalous in anomaly-based ID and intrusive/non-intrusive in misuse-based ID) and attain quite accurate results. However, the need for data labelling in the set-up phase and the continuous evolution of attack types often lead to a very expensive training process.

Human beings appear to be able to learn without explicit supervision. One aim of unsupervised learning is to mimic this aspect of human learning and hence this type of learning tends to use more biologically plausible methods than those using error descent methods. Unsupervised learning in ANNs meets the ID requirements as in a real-life situation there is no target reference with which to compare the response of the network. On the ID battlefield, there is no guarantee that a dataset labelled as "normal" will not contain any attack. Equally, a dataset that contains a fixed number of attacks can also contain new (unknown) attacks.

Usually, time information is not used in ANN-based IDSs. On the contrary, it can be employed as one of the inputs to the neural model embedded in MOVICAB-IDS to show how the traffic data evolve. It helps to identify anomalous situations by taking into account such aspects as high packet density and temporal evolution in non-parallel directions. By retaining temporal information, the context of intrusions is also kept, which is an important feature in visually analysing traffic data [172]. Thus, analysts can gain key knowledge about the nature of the anomalous situations to decide whether they are malignant. This is an important issue in ID because, to properly diagnose an event, analysts must assemble the contextual information of an attack [172].

Knowledge discovery, pattern recognition, data mining, and other such techniques, deal with the problem of extracting interesting classifications, clusters, associations, and other patterns from data. The existence of laptops, handhelds, embedded systems, and wearable computers is making ubiquitous access to a large quantity of distributed data a reality. Advanced analysis of distributed data for extracting useful knowledge is the next natural step in the increasingly interconnected world of ubiquitous and distributed computing. MOVICAB-IDS has been designed to make it accessible from any mobile device enabling permanent mobile visualisation, monitoring, and supervision of computer networks.

Considering all of the above mentioned issues, it was decided to design a hybrid visualisation-based IDS. The design of the system, employed different AI paradigms, involving the following advantages:

- **Generalization capability:** by using the CMLHL model, previously unseen attacks may be identified.
- **Scalability:** new agents (both SNIFFER and ANALYSER) can be dynamically added at any time.
- **Failure tolerance:** backup instances of some agents can be ready to run as soon as the working instances fail, showing a proactive behaviour.
- **Resource optimisation:** thanks to the CBP-BDI agents, the heterogeneous computing resources of a network can be efficiently used for the data analysis task.
- **Execution-time processing:** by splitting the data and allowing the system to analyse it in different processing units (agents located in different machines).
- **Mobile visualisation:** the visualisation task can be performed in a wide variety of devices.

On the other hand, some limitations of the proposed model have been also identified:

- **Human processing:** as it is based on unsupervised learning, MOVICAB-IDS can not automatically raise an alarm to warn about attacks. Hence, human supervision is required to identify the anomalous situations. Additionally, human users can fail to detect an intrusion even when visualised as an anomalous one.
- **Limited detection:** due to the reduced feature set used for packet projection, some attacks (such as those contained in the packets payload) can not be identified by means of MOVICAB-IDS.
- **Self protecting:** as pointed out in [45], if you use a security visualisation system, sooner or later it will be attacked. Compromising a security visualisation system is a new challenge for attackers. Although some security mechanisms have been implemented in MOVICAB-IDS (replicated agents and replanning mechanisms of the COORDINATOR agent), it performs poorly when defending itself from attacks.

Finally, MOVICAB-IDS is proposed as complementary to other network security tools. A recent trend in ID, named IDS cooperation, is to use different IDSs at the same time and correlate the analysis results and corresponding alerts [41]. MOVICAB-IDS can work in unison with other IDSs by using them in parallel on the same data source. A similar idea is followed in [166]. IDS cooperation can improve the ID results in several ways:

- MOVICAB-IDS visualises some attacks in a similar way. By combining it with some other IDSs, it will be possible to discriminate among them.
- By combining MOVICAB-IDS with a misuse-based classification IDS, MOVICAB-IDS can help in identifying 0-day attacks, which will allow the MIDS attack signatures to be updated.

7.3 Future Work

After identifying the main weaknesses of MOVICAB-IDS, future work will focus on the following areas:

- MOVICAB-IDS has been designed to identify SNMP-related attacks successfully. Unfortunately, there are plenty of other attacks that this IDS could target.
- Improvements to the dynamic assignment of analysis in order to employ computational resources in more efficient ways. Additionally, the COORDINATOR agent should be upgraded to deal properly with a high performance computing cluster.
- The testing and assessment of IDSs is still an open field, as pointed out in Chapter 6. This is especially noticeable in the case of visualisation-based IDSs. Nevertheless, a methodology to compare visualisation-based IDSs employing different visualisation techniques has yet to be developed.
- As stated in [324], "*since an agent is easily subverted by a process that is faulty, a multi-agent based intrusion detection system must be fault tolerant by being able to recover from system crashes, caused either accidentally or by malicious activity*". Thus, securing MOVICAB-IDS components remains pending. Although, as previously stated, some security mechanisms have been included in MOVICAB-IDS, it is not fully resistant to every single attack.
- There is a need to complement MOVICAB-IDS facilities to integrate some other components such as alert correlation and packet exploration (different levels of abstraction). This is in line with Shneiderman's information visualisation mantra [325]: "*overview first, zoom and filter, then details on demand*". Incorporating other data views will make it all the more difficult to evade this visualisation tool.

References

1. Lippmann, R.P., Fried, D.J., Graf, I., Haines, J.W., Kendall, K.R., McClung, D., Weber, D., Webster, S.E., Wyschogrod, D., Cunningham, R.K., Zissman, M.A.: Evaluating Intrusion Detection Systems: the 1998 DARPA Off-line Intrusion Detection Evaluation. In: DARPA Information Survivability Conference and Exposition, DISCEX 2000 (2000)
2. CERT Statistics (Historical), `http://www.cert.org/stats/`
3. Internet Security Threat Report - Volume XIV - Executive Summary. Symantec Corporation (2009)
4. Richardson, R.: CSI Survey 2008 - The 13th Annual Computer Crime & Security Survey. Computer Security Institute (2008)
5. Richardson, R.: CSI Survey 2007 - The 12th Annual Computer Crime and Security Survey. Computer Security Institute (2007)
6. X-Force Threat Insight Quarterly - Q4 2008, IBM Internet Security Systems - X-Force (2009)
7. Galetsas, A.: Statistical Data on Network Security. European Commission. Information Society and Media Directorate-General. Emerging Technologies & Infrastructures. Security (2007)
8. Barbará, D., Jajodia, S.: Applications of Data Mining in Computer Security. In: Advances in Information Security, vol. 6. Kluwer Academic Publishers, Dordrecht (2002)
9. Neumann, P.G., Parker, D.B.: A Summary of Computer Misuse Techniques. In: 12th National Computer Security Conference (1989)
10. Rogers, L.R.: Home Computer and Internet User Security. CERT, Software Engineering Institute, Carnegie Mellon University (2005)
11. Goodall, J.R., Lutters, W.G., Komlodi, A.: The Work of Intrusion Detection: Rethinking the Role of Security Analysts. In: Americas Conference on Information Systems (2004)
12. Heady, R., Luger, G., Maccabe, A., Servilla, M.: The Architecture of a Network Level Intrusion Detection System. Technical Report CS90-20. University of New Mexico (1990)
13. Security Terms, `https://security.ias.edu/glossary`
14. Puketza, N.J., Zhang, K., Chung, M., Mukherjee, B., Olsson, R.A.: A Methodology for Testing Intrusion Detection Systems. IEEE Transactions on Software Engineering 22(10), 719–729 (1996)
15. Woon, I.M.Y., Kankanhalli, A.: Investigation of IS Professionals' Intention to Practise Secure Development of Applications. International Journal of Human-Computer Studies 65(1), 29–41 (2007)

16. Allen, J., Christie, A., Fithen, W., McHugh, J., Pickel, J., Stoner, E.: State of the Practice of Intrusion Detection Technologies. Technical Report CMU/SEI-99-TR-028. Carnegie Mellon University - Software Engineering Institute (2000)
17. Bace, R., Mell, P.: NIST Special Publication on Intrusion Detection Systems. National Institute of Standards and Technology - U.S. Department of Commerce (2001)
18. X-Force 2008, Trend & Risk Report. IBM Internet Security Systems - X-Force (2009)
19. Maloof, M.: Machine Learning and Data Mining for Computer Security. In: Advanced Information and Knowledge Processing. Springer, Heidelberg (2006)
20. Chuvakin, A.: Monitoring IDS. Information Security Journal: A Global Perspective 12(6), 12–16 (2004)
21. Rizza, J.M.: Computer Network Security. Springer, US (2005)
22. Mukherjee, B., Heberlein, L.T., Levitt, K.N.: Network Intrusion Detection. IEEE Network 8(3), 26–41 (1994)
23. Marchette, D.J.: Computer Intrusion Detection and Network Monitoring: A Statistical Viewpoint. Information Science and Statistics Springer-Verlag New York, Inc., New York (2001)
24. McHugh, J., Christie, A., Allen, J.: Defending Yourself: The Role of Intrusion Detection Systems. IEEE Software 17(5), 42–51 (2000)
25. Smaha, S.E.: Haystack: an Intrusion Detection System. In: Fourth Aerospace Computer Security Applications Conference (1988)
26. Lindqvist, U., Jonsson, E.: How to Systematically Classify Computer Security Intrusions. In: 1997 IEEE Symposium on Security and Privacy (1997)
27. Khaled, L.: Computer Security and Intrusion Detection. Crossroads 11(1), 2 (2004)
28. Anderson, J.P.: Computer Security Threat Monitoring and Surveillance. Technical Report (1980)
29. Denning, D.E.: An Intrusion-Detection Model. IEEE Transactions on Software Engineering 13(2), 222–232 (1987)
30. Jones, A., Sielken, R.: Computer System Intrusion Detection: A Survey. White Paper. University of Virginia - Computer Science Department (1999)
31. McHugh, J.: Intrusion and Intrusion Detection. International Journal of Information Security 1(1), 14–35 (2001)
32. Debar, H., Dacier, M., Wespi, A.: Towards a Taxonomy of Intrusion-Detection Systems. Computer Networks - the International Journal of Computer and Telecommunications Networking 31(8), 805–822 (1999)
33. Ptacek, T.H., Newsham, T.N.: Insertion, Evasion, and Denial of Service: Eluding Network Intrusion Detection. White paper. Secure Networks, Inc. (1998)
34. Handley, M., Paxson, V., Kreibich, C.: Network Intrusion Detection: Evasion, Traffic Normalization, and End-to-end Protocol Semantics. In: 10th USENIX Security Symposium (2001)
35. Lazarevic, A., Kumar, V., Srivastava, J.: Intrusion Detection: a Survey. Managing Cyber Threats: Issues, Approaches, and Challenges. Massive Computing 5, 19–78 (2005)
36. Sundaram, A.: An Introduction to Intrusion Detection. Crossroads 2(4), 3–7 (1996)
37. Laskov, P., Düssel, P., Schäfer, C., Rieck, K.: Learning intrusion detection: Supervised or unsupervised? In: Roli, F., Vitulano, S. (eds.) ICIAP 2005. LNCS, vol. 3617, pp. 50–57. Springer, Heidelberg (2005)

38. Balepin, I., Maltsev, S., Rowe, J., Levitt, K.N.: Using specification-based intrusion detection for automated response. In: Vigna, G., Krügel, C., Jonsson, E. (eds.) RAID 2003. LNCS, vol. 2820, pp. 136–154. Springer, Heidelberg (2003)
39. Kemmerer, R.A., Vigna, G.: Intrusion Detection: a Brief History and Overview. Computer 35(4), 27–30 (2002)
40. Verwoerd, T., Hunt, R.: Intrusion Detection Techniques and Approaches. Computer Communications 25(15), 1356–1365 (2002)
41. Kruegel, C., Valeur, F., Vigna, G.: Intrusion Detection and Correlation - Challenges and Solutions. In: Advances in Information Security. Springer, US (2005)
42. Ahlberg, C., Shneiderman, B.: Visual Information Seeking: Tight Coupling of Dynamic Query Filters with Starfield Displays. In: Readings in Information Visualization: using Vision to Think, pp. 244–250. Morgan Kaufmann Publishrs Inc., San Francisco (1999)
43. Becker, R.A., Eick, S.G., Wilks, A.R.: Visualizing Network Data. IEEE Transactions on Visualization and Computer Graphics 1(1), 16–28 (1995)
44. Choi, H., Lee, H., Kim, H.: Fast Detection and Visualization of Network Attacks on Parallel Coordinates. Computers & Security 28(5), 276–288 (2009)
45. Conti, G.: Security Data Visualization: Graphical Techniques for Network Analysis. No Starch Press (2007)
46. Marty, R.: Applied Security Visualization. Addison-Wesley Professional, Reading (2008)
47. Itoh, T., Takakura, H., Sawada, A., Koyamada, K.: Hierarchical Visualization of Network Intrusion Detection Data. IEEE Computer Graphics and Applications 26(2), 40–47 (2006)
48. Conti, G., Abdullah, K.: Passive Visual Fingerprinting of Network Attack Tools. In: 2004 ACM Workshop on Visualization and Data Mining for Computer Security, ACM, Washington (2004)
49. Wilkinson, L.: The Grammar of Graphics. In: Statistics and Computing. Springer, New York (2005)
50. Keim, D.A.: Information Visualization and Visual Data Mining. IEEE Transactions on Visualization and Computer Graphics 8(1), 1–8 (2002)
51. Elmqvist, N., Dragicevic, P., Fekete, J.D.: Rolling the Dice: Multidimensional Visual Exploration using Scatterplot Matrix Navigation. IEEE Transactions on Visualization and Computer Graphics 14(6), 1141–1148 (2008)
52. Iris Dataset - UCI Machine Learning Repository, `http://archive.ics.uci.edu/ml/datasets/Iris`
53. McCulloch, W.S., Pitts, W.: A Logical Calculus of the Ideas Immanent in Nervous Activity. Bulletin of Mathematical Biology 5(4), 115–133 (1943)
54. Hebb, D.: The Organisation of Behaviour. Willey, New York (1949)
55. Haykin, S.: Neural Networks: a Comprehensive Foundation. Macmillan, Basingstoke (1994)
56. Rubner, J., Tavan, P.: A Self-Organizing Network for Principal Component Analysis. Europhysics Letters 10(7), 693–698 (1989)
57. Rubner, J., Schulten, K.: Development of Feature Detectors by Self-Organization. Biological Cybernetics 62(3), 193–199 (1990)
58. Rumelhart, D.E., Zipser, D.: Feature Discovery by Competitive Learning. Parallel Distributed Processing: Explorations in the Microstructure of Cognition. Foundations, vol. 1. MIT Press, Cambridge (1986)

59. Pearson, K.: On Lines and Planes of Closest Fit to Systems of Points in Space. Philosophical Magazine 2(6), 559–572 (1901)
60. Hotelling, H.: Analysis of a Complex of Statistical Variables into Principal Components. Journal of Education Psychology 24, 417–444 (1933)
61. Bishop, C.M.: Neural Networks for Pattern Recognition. Oxford University Press, Oxford (1996)
62. Oja, E.: Neural Networks, Principal Components, and Subspaces. International Journal of Neural Systems 1, 61–68 (1989)
63. Oja, E.: Principal Components, Minor Components, and Linear Neural Networks. Neural Networks 5(6), 927–935 (1992)
64. Oja, E.: A Simplified Neuron Model as a Principal Component Analyzer. Journal of Mathematical Biology 15(3), 267–273 (1982)
65. Sanger, D.: Contribution Analysis: a Technique for Assigning Responsibilities to Hidden Units in Connectionist Networks. Connection Science 1(2), 115–138 (1989)
66. Fyfe, C.: A Neural Network for PCA and Beyond. Neural Processing Letters 6(1-2), 33–41 (1997)
67. Oja, E., Ogawa, H., Wangviwattana, J.: Principal Component Analysis by Homogeneous Neural Networks, Part I: The Weighted Subspace Criterion. IEICE Transactions on Information and Systems 75(3), 366–375 (1992)
68. Fyfe, C.: PCA Properties of Interneurons: from Neurobiology to Real World Computing. In: International Conference on Artificial Neural Networks (ICANN 1993). Springer, Heidelberg (1993)
69. Fyfe, C.: Negative Feedback as an Organising Principle for Artificial Neural Networks. PhD Thesis. Strathclyde University (1995)
70. Charles, D., Fyfe, C.: Modelling Multiple-Cause Structure using Rectification Constraints. Network: Computation in Neural Systems 9(2), 167–182 (1998)
71. Fyfe, C., Corchado, E.: Maximum Likelihood Hebbian Rules. In: 10th European Symposium on Artificial Neural Networks, ESANN 2002 (2002)
72. Karhunen, J., Joutsensalo, J.: Representation and Separation of Signals using Nonlinear PCA Type Learning. Neural Networks 7(1), 113–127 (1994)
73. Oja, E., Ogawa, H., Wangviwattana, J.: Learning in Nonlinear Constrained Hebbian Networks. In: International Conference on Artificial Neural Networks (1991)
74. Xu, L.: Least Mean Square Error Reconstruction Principle for Self-organizing Neural-nets. Neural Networks 6(5), 627–648 (1993)
75. Bell, A.J., Sejnowski, T.J.: An Information-Maximization Approach to Blind Separation and Blind Deconvolution. Neural Computation 7(6), 1129–1159 (1995)
76. Girolami, M., Fyfe, C.: Stochastic ICA Contrast Maximisation using OJA's Nonlinear PCA Algorithm. International Journal of Neural Systems 8(5-6), 661 (1999)
77. Friedman, J.H., Tukey, J.W.: A Projection Pursuit Algorithm for Exploratory Data-Analysis. IEEE Transactions on Computers 23(9), 881–890 (1974)
78. Diaconis, P., Freedman, D.: Asymptotics of Graphical Projection Pursuit. The Annals of Statistics 12(3), 793–815 (1984)
79. Fyfe, C., Baddeley, R., McGregor, D.R.: Exploratory Projection Pursuit: an Artificial Neural Network Approach. Research Report/94/160, University of Strathclyde (1994)
80. Corchado, E., MacDonald, D., Fyfe, C.: Maximum and Minimum Likelihood Hebbian Learning for Exploratory Projection Pursuit. Data Mining and Knowledge Discovery 8(3), 203–225 (2004)

81. Corchado, E., Fyfe, C.: Connectionist Techniques for the Identification and Suppression of Interfering Underlying Factors. International Journal of Pattern Recognition and Artificial Intelligence 17(8), 1447–1466 (2003)
82. Corchado, E., Han, Y., Fyfe, C.: Structuring Global Responses of Local Filters Using Lateral Connections. Journal of Experimental & Theoretical Artificial Intelligence 15(4), 473–487 (2003)
83. Seung, H.S., Socci, N.D., Lee, D.: The Rectified Gaussian Distribution. Advances in Neural Information Processing Systems 10, 350–356 (1998)
84. Corchado, E., Burgos, P., del Mar Rodríguez, M., Tricio, V.: A Hierarchical Visualization Tool to Analyse the Thermal Evolution of Construction Materials. In: Luo, Y. (ed.) CDVE 2004. LNCS, vol. 3190, pp. 238–245. Springer, Heidelberg (2004)
85. Corchado, E., Pellicer, M.A., Borrajo, M.L.: A MLHL Based Method to an Agent-Based Architecture. International Journal of Computer Mathematics (2009) (accepted in press)
86. Herrero, Á., Corchado, E., Sáiz, L., Abraham, A.: DIPKIP: A Connectionist Knowledge Management System to Identify Knowledge Deficits in Practical Cases. Computational Intelligence (2009) (accepted in press)
87. Kohonen, T.: The Self-Organizing Map. IEEE 78(9), 1464–1480 (1990)
88. Ritter, H., Martinetz, T., Schulten, K.: Neural Computation and Self-Organizing Maps; An Introduction. Addison-Wesley Longman Publishing Co., Inc., Amsterdam (1992)
89. Oja, M., Kaski, S., Kohonen, T.: Bibliography of Self-Organizing Map (SOM) Papers: 1998-2001 Addendum. Neural Computing Surveys 3(1), 1–156 (2003)
90. Demartines, P., Herault, J.: Curvilinear Component Analysis: A Self-Organizing Neural Network for Nonlinear Mapping of Data Sets. IEEE Transactions on Neural Networks 8(1), 148–154 (1997)
91. Demartines, P.: Analyze de données par réseaux de neurones auto-organizés. Institut National Polytechnique de Grenoble (1994)
92. Franklin, S., Graesser, A.: Is It an Agent, or Just a Program?: A Taxonomy for Autonomous Agents. In: Jennings, N.R., Wooldridge, M.J., Müller, J.P. (eds.) ECAI-WS 1996 and ATAL 1996. LNCS, vol. 1193. Springer, Heidelberg (1997)
93. Russell, S.J., Norvig, P.: Artificial Intelligence: a Modern Approach. Prentice Hall, Englewood Cliffs (1995)
94. Weiss, G.: Multiagent Systems: a Modern Approach to Distributed Artificial Intelligence. MIT Press, Cambridge (1999)
95. Ferber, J.: Multi-agent Systems: an Introduction to Distributed Artificial Intelligence. Addison-Wesley, Reading (1999)
96. Huhns, M.N., Singh, M.P.: A Multiagent Treatment of Agenthood. Applied Artificial Intelligence 13(1-2), 3–10 (1999)
97. Durfee, E.H., Lesser, V.R.: Negotiating Task Decomposition and Allocation Using Partial Global Planning. Distributed Artificial Intelligence, vol. 2. Morgan Kaufmann Publishers Inc., San Francisco (1989)
98. Jennings, N.R., Sycara, K., Wooldridge, M.: A Roadmap of Agent Research and Development. Autonomous Agents and Multi-Agent Systems 1(1), 7–38 (1998)
99. Wooldridge, M.: Agent-based Computing. Interoperable Communication Networks 1(1), 71–97 (1998)
100. Brenner, W., Wittig, H., Zarnekow, R.: Intelligent Software Agents: Foundations and Applications. Springer-Verlag New York, Inc., Secaucus (1998)

101. Bird, S.D.: Toward a Taxonomy of Multi-agent Systems. International Journal of Man-Machine Studies 39(4), 689–704 (1993)
102. Wooldridge, M., Jennings, N.R.: Intelligent Agents: Theory and Practice. Knowledge Engineering Review 10(2), 115–152 (1995)
103. Leake, D., Wilson, D.: When experience is wrong: Examining CBR for changing tasks and environments. In: Althoff, K.-D., Bergmann, R., Branting, L.K. (eds.) ICCBR 1999. LNCS (LNAI), vol. 1650, p. 218. Springer, Heidelberg (1999)
104. Aamodt, A., Plaza, E.: Case-Based Reasoning - Foundational Issues, Methodological Variations, and System Approaches. AI Communications 7(1), 39–59 (1994)
105. Mantaras, R.L.D., McSherry, D., Bridge, D., Leake, D., Smyth, B., Craw, S., Faltings, B., Maher, M.L., Cox, M.T., Forbus, K., Keane, M., Aamodt, A., Watson, I.: Retrieval, Reuse, Revision, and Retention in Case-Based Reasoning. The Knowledge Engineering Review 20(3), 215–240 (2005)
106. Kolodner, J.: Case-based Reasoning. Morgan Kaufmann Publishers Inc., San Francisco (1993)
107. Reinartz, T., Iglezakis, I., Berghofer, T.R.: Review and Restore for Case-Base Maintenance. Computational Intelligence 17(2), 214–234 (2001)
108. Eskin, E., Arnold, A., Prerau, M., Portnoy, L., Stolfo, S.: A Geometric Framework for Unsupervised Anomaly Detection: Detecting Intrusions in Unlabeled Data. In: Barbará, D., Jajodia, S. (eds.) Applications of Data Mining in Computer Security, pp. 77–101. Kluwer, Dordrecht (2002)
109. SNORT - Open Source Network Intrusion Prevention and Detection System, `http://snort.org/`
110. Noel, S., Wijesekera, D.: Modern Intrusion Detection, Data Mining, and Degrees of Attack Guilt. In: Barbará, D., Jajodia, S. (eds.) Applications of Data Mining in Computer Security, pp. 1–31. Kluwer Academic Publishers, Dordrecht (2002)
111. Julisch, K.: Data Mining for Intrusion Detection: A Critical Review. In: Barbará, D., Jajodia, S. (eds.) Applications of Data Mining in Computer Security. Advances in Information Security, pp. 33–62. Kluwer Academic Publishers, Dordrecht (2002)
112. Ye, N., Chen, Q.: Attack-norm Separation for Detecting Attack-induced Quality Problems on Computers and Networks. Quality and Reliability Engineering International 23(5), 545–553 (2007)
113. Eleazar, E.: Anomaly Detection over Noisy Data using Learned Probability Distributions. In: Seventeenth International Conference on Machine Learning. Morgan Kaufmann Publishers Inc., San Francisco (2000)
114. Lane, T., Brodley, C.E.: Temporal Sequence Learning and Data Reduction for Anomaly Detection. ACM Transactions on Information and System Security 2(3), 295–331 (1999)
115. Katos, V.: Network Intrusion Detection: Evaluating Cluster, Discriminant, and Logit Analysis. Information Sciences 177(15), 3060–3073 (2007)
116. Sarasamma, S.T., Zhu, Q.A.: Min-Max Hyperellipsoidal Clustering for Anomaly Detection in Network Security. IEEE Transactions on Systems, Man, and Cybernetics, Part B: Cybernetics 36(4), 887–901 (2006)
117. Wu, N., Zhang, J.: Factor-Analysis Based Anomaly Detection and Clustering. Decision Support Systems 42(1), 375–389 (2006)
118. Engelhardt, D.: Directions for Intrusion Detection and Response: a Survey. Electronics and Surveillance Research Laboratory, Defence Science and Technology Organisation, Department of Defence, Australian Government (1997)

119. Blustein, J., Fu, C.L., Silver, D.L.: Information Visualization for an Intrusion Detection System. In: Sixteenth ACM Conference on Hypertext and Hypermedia. ACM Press, New York (2005)
120. Goodall, J.R.: User Requirements and Design of a Visualization for Intrusion Detection Analysis. In: Sixth Annual IEEE SMC Information Assurance Workshop, IAW 2005 (2005)
121. Withall, M., Phillips, I., Parish, D.: Network Visualisation: a Review. IET Communications 1(3), 365–372 (2007)
122. Pongsiri, J., Parikh, M., Raspopovic, M., Chandra, K.: Visualization of Internet Traffic Features. In: 12th International Conference of Scientific Computing and Mathematical Modeling (1999)
123. Cox, K.C., Eick, S.G., He, T.: 3D Geographic Network Displays. ACM SIGMOD Record 25(4), 50–54 (1996)
124. Koutsofios, E.E., North, S.C., Truscott, R., Keim, D.A.: Visualizing Large-Scale Telecommunication Networks and Services. In: Conference on Visualization 1999. IEEE Computer Society Press, San Francisco (1999)
125. Dodge, M., Kitchin, R.: Atlas of Cyberspace. Addison-Wesley, Reading (2001)
126. Mansmann, F., Keim, D.A., North, S.C., Rexroad, B., Sheleheda, D.: Visual Analysis of Network Traffic for Resource Planning, Interactive Monitoring, and Interpretation of Security Threats. IEEE Transactions on Visualization and Computer Graphics 13(6), 1105–1112 (2007)
127. Claffy, K., Hyun, Y., Keys, K., Fomenkov, M., Krioukov, D.: Internet Mapping: From Art to Science. In: Conference For Homeland Security, Cybersecurity Applications & Technology (2009)
128. MRTG: The Multi Router Traffic Grapher, `http://www.mrtg.org`
129. Nyarko, K., Capers, T., Scott, C., Ladeji-Osias, K.A.: Network Intrusion Visualization with NIVA, an Intrusion Detection Visual Analyzer with Haptic Integration. In: Capers, T. (ed.) 10th Symposium on Haptic Interfaces for Virtual Environment and Teleoperator Systems, HAPTICS 2002 (2002)
130. Wood, A., Mountain, S.R., Denver, C.O., Practical, G.G.: Intrusion Detection: Visualizing Attacks in IDS Data. GIAC GCIA Practical, SANS Institute (2003)
131. Koike, H., Ohno, K.: SnortView: Visualization System of Snort Logs. In: 2004 ACM Workshop on Visualization and Data Mining for Computer Security. ACM Press, Washington (2004)
132. Abdullah, K., Lee, C.P., Conti, G., Copeland, J.A., Stasko, J.: IDS RainStorm: Visualizing IDS Alarms. In: IEEE Workshop on Visualization for Computer Security (VizSEC 2005). IEEE Computer Society, Los Alamitos (2005)
133. Koike, H., Ohno, K., Koizumi, K.: Visualizing Cyber Attacks Using IP Matrix. In: IEEE Workshop on Visualization for Computer Security (VizSEC 2005). IEEE Computer Society, Los Alamitos (2005)
134. Axelsson, S., Sands, D.: Understanding Intrusion Detection through Visualization. In: Advances in Information Security, vol. 24. Springer, Heidelberg (2006)
135. Foresti, S., Agutter, J., Livnat, Y., Moon, S., Erbacher, R.: Visual Correlation of Network Alerts. IEEE Computer Graphics and Applications 26(2), 48–59 (2006)
136. Analysis Console for Intrusion Databases (ACID), `http://www.andrew.cmu.edu/user/rdanyliw/snort/snortacid.html`
137. Kim, H., Kang, I., Bahk, S.: Real-time Visualization of Network Attacks on High-speed Links. IEEE Network 18(5), 30–39 (2004)

138. Lau, S.: The Spinning Cube of Potential Doom. Communications of the ACM 47(6), 25–26 (2004)
139. The Shoki Packet Hustler, http://shoki.sourceforge.net/hustler/
140. Plonka, D.: FlowScan: A Network Traffic Flow Reporting and Visualization Tool. In: 14th USENIX Conference on System Administration. USENIX Association, New Orleans (2000)
141. Abdullah, K., Lee, C., Conti, G., Copeland, J.A.: Visualizing Network Data for Intrusion Detection. In: Sixth Annual IEEE Information Assurance Workshop - Systems, Man and Cybernetics (2005)
142. Stockinger, K., Bethel, E.W., Campbell, S., Dart, E., Wu, K.: Detecting Distributed Scans using High-Performance Query-Driven Visualization. In: 2006 ACM/ IEEE Conference on Supercomputing. ACM, Tampa (2006)
143. Conti, G., Grizzard, J., Ahamad, M., Owen, H.: Visual Exploration of Malicious Network Objects Using Semantic Zoom, Interactive Encoding and Dynamic Queries. In: IEEE Workshop on Visualization for Computer Security (VizSEC 2005). IEEE Computer Society, Los Alamitos (2005)
144. Taylor, T., Brooks, S., McHugh, J.: NetBytes Viewer: An Entity-Based NetFlow Visualization Utility for Identifying Intrusive Behavior. In: Workshop on Visualization for Computer Security (VizSEC 2007). Mathematics and Visualization. Springer, Heidelberg (2008)
145. Cheung, S., Crawford, R., Dilger, M., Frank, J., Hoagland, J., Levitt, K., Rowe, J., Staniford-Chen, S., Yip, R., Zerkle, D.: The Design of GrIDS: A Graph-based Intrusion Detection System. Technical Report CSE-99-2. University of California at Davis. Computer Science Department (1999)
146. Yin, X., Yurcik, W., Adam, S.: VisFlowCluster-IP: Connectivity-Based Visual Clustering of Network Hosts. In: 21st IFIP International Information Security Conference (SEC 2006), Karlstad, Sweden (2006)
147. Pearlman, J., Rheingans, P.: Visualizing Network Security Events Using Compound Glyphs from a Service-Oriented Perspective. In: Workshop on Visualization for Computer Security (VizSEC 2007). Mathematics and Visualization. Springer, Heidelberg (2008)
148. Höglund, A.J., Hätönen, K.: Computer Network User Behaviour Visualisation using Self Organizing Maps. In: 8th International Conference on Artificial Neural Networks (ICANN 1998), vol. 2. IEEE, Los Alamitos (1998)
149. Girardin, L.: An Eye on Network Intruder-Administrator Shootouts. In: 1st Workshop on Intrusion Detection and Network Monitoring 1. USENIX Association, Santa Clara (1999)
150. Girardin, L., Brodbeck, D.: A Visual Approach for Monitoring Logs. In: 12th USENIX Conference on System Administration. USENIX Association, Boston (1998)
151. Labib, K., Vemuri, V.: Application of Exploratory Multivariate Analysis for Network Security. In: Vemuri, V. (ed.) Enhancing Computer Security with Smart Technology, pp. 229–261. CRC Press, Boca Raton (2005)
152. Kayacik, H.G., Zincir-Heywood, A.N.: Using Self-Organizing Maps to Build an Attack Map for Forensic Analysis. In: ACM 2006 International Conference on Privacy, Security and Trust (PST 2006). ACM, Markham (2006)
153. KDD Cup 1999 Dataset (1999), http://archive.ics.uci.edu/ml/databases/kddcup99/kddcup99.html

154. Elkan, M.: Results of the KDD 1999 Classifier Learning Contest (1999), http://www-cse.ucsd.edu/users/elkan/clresults.html
155. Kumar, G., Devaraj, D.: Network Intrusion Detection using Hybrid Neural Networks. In: International Conference on Signal Processing, Communications and Networking, ICSCN 2007 (2007)
156. Solka, J.L., Marchette, D.J., Wallet, B.C.: Statistical Visualization Methods in Intrusion Detection. In: 32nd Symposium on the Interface. Computing Science and Statistics, vol. 32 (2000)
157. Estrin, D., Handley, M., Heidemann, J., McCanne, S., Ya, X., Yu, H.: Network Visualization with Nam, the VINT Network Animator. IEEE Computer Magazine 33(11), 63–68 (2000)
158. D'Amico, A., Larkin, M.: Methods of Visualizing Temporal Patterns in and Mission Impact of Computer Security Breaches. In: DARPA Information Survivability Conference & Exposition II (DISCEX 2001) (January 2001)
159. Erbacher, R.F., Frincke, D.: Visual Behavior Characterization for Intrusion and Misuse Detection. In: Visual Data Exploration and Analysis Conference. SPIE Proceedings Series, vol. 4302 (2001)
160. Erbacher, R.F., Garber, M.: Visualization Techniques for Intrusion Behavior Identification. In: Sixth Annual IEEE SMC Information Assurance Workshop, IAW 2005 (2005)
161. Teoh, S.T., Ma, K.-L., Wu, S.F., Jankun-Kelly, T.J.: Detecting Flaws and Intruders with Visual Data Analysis. IEEE Computer Graphics and Applications 24(5), 27–35 (2004)
162. McPherson, J., Ma, K.-L., Krystosk, P., Bartoletti, T., Christensen, M.: PortVis: a Tool for Port-Based Detection of Security Events. In: 2004 ACM Workshop on Visualization and Data Mining for Computer Security. ACM Press, Washington (2004)
163. Lakkaraju, K., Yurcik, W., Lee, A.J.: NVisionIP: Netflow Visualizations of System State for Security Situational Awareness. In: 2004 ACM Workshop on Visualization and Data Mining for Computer Security. ACM, Washington (2004)
164. Abad, C., Li, Y., Lakkaraju, K., Yin, X., Yurcik, W.: Correlation Between NetFlow System and Network Views for Intrusion Detection. In: Workshop on Link Analysis, Counter-terrorism, and Privacy. Lake Buena Vista, FL (2004)
165. Lee, C.P., Trost, J., Gibbs, N., Raheem, B., Copeland, J.A.: Visual Firewall: Real-time Network Security Monitor. In: IEEE Workshop on Visualization for Computer Security (VizSEC 2005). IEEE Computer Society, Los Alamitos (2005)
166. Oline, A., Reiners, D.: Exploring Three-Dimensional Visualization for Intrusion Detection. In: IEEE Workshop on Visualization for Computer Security (VizSEC 2005). IEEE Computer Society, Los Alamitos (2005)
167. Muelder, C., Ma, K.-L., Bartoletti, T.: A Visualization Methodology for Characterization of Network Scans. In: IEEE Workshop on Visualization for Computer Security (VizSEC 2005), IEEE Computer Society, Los Alamitos (2005)
168. Krasser, S., Conti, G., Grizzard, J., Gribschaw, J., Owen, H.: Real-time and Forensic Network Data Analysis Using Animated and Coordinated Visualization. In: Sixth Annual IEEE SMC Information Assurance Workshop, IAW 2005 (2005)
169. Fink, G.A., Muessig, P., North, C.: Visual Correlation of Host Processes and Network Traffic. In: IEEE Workshop on Visualization for Computer Security (VizSEC 2005). IEEE Computer Society, Los Alamitos (2005)

170. Ren, P., Gao, Y., Li, Z.C., Chen, Y., Watson, B.: IDGraphs: Intrusion Detection and Analysis Using Histographs. In: IEEE Workshop on Visualization for Computer Security (VizSEC 2005). IEEE Computer Society, Los Alamitos (2005)
171. Ren, P., Gao, Y., Li, Z.C., Chen, Y., Watson, B.: IDGraphs: Intrusion Detection and Analysis Using Stream Compositing. IEEE Computer Graphics and Applications 26(2), 28–39 (2006)
172. Goodall, J.R., Lutters, W.G., Rheingans, P., Komlodi, A.: Focusing on Context in Network Traffic Analysis. IEEE Computer Graphics and Applications 26(2), 72–80 (2006)
173. Muelder, C., Ma, K.-L., Bartoletti, T.: Interactive visualization for network and port scan detection. In: Valdes, A., Zamboni, D. (eds.) RAID 2005. LNCS, vol. 3858, pp. 265–283. Springer, Heidelberg (2006)
174. D'Amico, A.D., Goodall, J.R., Tesone, D.R., Kopylec, J.K.: Visual Discovery in Computer Network Defense. IEEE Computer Graphics and Applications 27(5), 20–27 (2007)
175. Goodall, J.R., Tesone, D.R.: Visual Analytics for Network Flow Analysis. In: Cybersecurity Applications & Technology Conference for Homeland Security. IEEE Press, Los Alamitos (2009)
176. Fischer, F., Mansmann, F., Keim, D.A., Pietzko, S., Waldvogel, M.: Large-scale network monitoring for visual analysis of attacks. In: Goodall, J.R., Conti, G., Ma, K.-L. (eds.) VizSec 2008. LNCS, vol. 5210, pp. 111–118. Springer, Heidelberg (2008)
177. Shneiderman, B.: Tree Visualization with Tree-maps: a 2-D Space-filling Approach. ACM Transactions on Graphics 11(1), 92–99 (1992)
178. Phan, D., Gerth, J., Lee, M., Paepcke, A., Winograd, T.: Visual Analysis of Network Flow Data with Timelines and Event Plots. In: Workshop on Visualization for Computer Security (VizSEC 2007). Mathematics and Visualization. Springer, Heidelberg (2008)
179. Shoch, J.F., Hupp, J.A.: The "Worm" Programs - Early Experience with a Distributed Computation. Communications of the ACM 25(3), 172–180 (1982)
180. Crosbie, M., Dole, B., Ellis, T., Krsul, I., Spafford, E.: IDIOT Users Guide. Technical Report CSD-TR-96-050. Department of Computer Sciences. Purdue University (1996)
181. Jensen, K.: Coloured Petri Nets and the Invariant Method. Theoretical Computer Science 14, 317–336 (1981)
182. Petri, C.A.: Kommunikation mit Automaten. Institut für Instrumentelle Mathematik, Schriften des IMM Nr. 2 (1962)
183. Vert, G., Frincke, D.A., McConnell, J.C.: A Visual Mathematical Model for Intrusion Detection. In: 21st National Information Systems Security Conference (1998)
184. Atkison, T., Pensy, K., Nicholas, C., Ebert, D., Atkison, R., Morris, C.: Case Study: Visualization and Information Retrieval Techniques for Network Intrusion Detection. In: Joint Eurographics- IEEE TCVG Symposium on Visualization (VisSym 2001). Computer Science. Springer, Heidelberg (2001)
185. Ebert, D.S., Shaw, C.D., Zwa, A., Starr, C.: Two-handed Interactive Stereoscopic Visualization. In: IEEE 7th Conference on Visualization 1996. IEEE Computer Society Press, San Francisco (1996)
186. Takada, T., Koike, H.: Tudumi: Information Visualization System for Monitoring and Auditing Computer Logs. In: Sixth International Conference on Information Visualisation (2002)

187. Teoh, S.T., Ma, K.L., Wu, S.F., Zhao, X.: Case Study: Interactive Visualization for Internet Security. In: IEEE Conference on Visualization (Vis 2002). IEEE Computer Society, Boston (2002)
188. Fisk, M., Smith, S.A., Weber, P., Kothapally, S., Caudell, T.: Immersive Network Monitoring. In: 2003 Passive and Active Measurement Workshop (2003)
189. Ball, R., Glenn, A.F., North, C.: Home-centric Visualization of Network Traffic for Security Administration. In: 2004 ACM Workshop on Visualization and Data Mining for Computer Security. ACM, Washington (2004)
190. Erbacher, R.F., Christensen, K., Sundberg, A.: Designing Visualization Capabilities for IDS Challenges. In: IEEE Workshop on Visualization for Computer Security (VizSEC 2005). IEEE Computer Society, Los Alamitos (2005)
191. Komlodi, A., Rheingans, P., Utkarsha, A., Goodall, J.R., Amit, J.: A User-Centered Look at Glyph-Based Security Visualization. In: IEEE Workshop on Visualization for Computer Security (VizSEC 2005). IEEE Computer Society, Los Alamitos (2005)
192. Kim, S.S., Reddy, A.L.N.: A Study of Analyzing Network Traffic as Images in Real-time. In: IEEE 24th Annual Joint Conference of the IEEE Computer and Communications Societies, INFOCOM 2005 (March 2005)
193. Kisner, T., Essoh, A., Kaderali, F.: Visualisation of Network Traffic using Dynamic Co-occurrence Matrices. In: Second International Conference on Internet Monitoring and Protection, ICIMP 2007 (2007)
194. Onut, I.-V., Ghorbani, A.A.: SVision: A Novel Visual Network-Anomaly Identification Technique. Computers & Security 26(3), 201–212 (2007)
195. Samak, T., Ghanem, S., Ismail, M.A.: On the Efficiency of Using Space-filling Curves in Network Traffic Representation. In: IEEE INFOCOM Workshop (2008)
196. Paxson, V.: Bro: A System for Detecting Network Intruders in Real-Time. Computer Networks: the International Journal of Computer and Telecommunications Networking 31(23-24), 2435–2463 (1999)
197. Bro Intrusion Detection System, `http://bro-ids.org/`
198. tcpdump/libpcap, `http://www.tcpdump.org/`
199. Lippmann, R., Haines, J.W., Fried, D.J., Korba, J., Das, K.: The 1999 DARPA Off-line Intrusion Detection Evaluation. Computer Networks 34(4), 579–595 (2000)
200. Lippmann, R., Haines, J.W., Fried, D.J., Korba, J., Das, K.: Analysis and Results of the 1999 DARPA Off-Line Intrusion Detection Evaluation. In: Debar, H., Mé, L., Wu, S.F. (eds.) RAID 2000. LNCS, vol. 1907, p. 162. Springer, Heidelberg (2000)
201. Stolfo, S., Prodromidis, A.L., Tselepis, S., Lee, W., Fan, D.W., Chan, P.K.: JAM: Java Agents for Meta-Learning over Distributed Databases. In: Third International Conference on Knowledge Discovery and Data Mining (1997)
202. Reilly, M., Stillman, M.: Open Infrastructure for Scalable Intrusion Detection. In: 1998 IEEE Information Technology Conference (1998)
203. Spafford, E.H., Zamboni, D.: Intrusion Detection Using Autonomous Agents. Computer Networks: The International Journal of Computer and Telecommunications Networking 34(4), 547–570 (2000)
204. Hegazy, I.M., Al-Arif, T., Fayed, Z.T., Faheem, H.M.: A Multi-agent Based System for Intrusion Detection. IEEE Potentials 22(4), 28–31 (2003)
205. Gorodetski, V., Kotenko, I., Karsaev, O.: Multi-Agent Technologies for Computer Network Security: Attack Simulation, Intrusion Detection and Intrusion Detection Learning. Computer Systems Science and Engineering 18(4), 191–200 (2003)

206. Miller, P., Inoue, A.: Collaborative Intrusion Detection System. In: 22nd International Conference of the North American Fuzzy Information Processing Society, NAFIPS 2003 (2003)
207. Gorodetsky, V., Karsaev, O., Samoilov, V., Ulanov, A.: Asynchronous alert correlation in multi-agent intrusion detection systems. In: Gorodetsky, V., Kotenko, I., Skormin, V.A. (eds.) MMM-ACNS 2005. LNCS, vol. 3685, pp. 366–379. Springer, Heidelberg (2005)
208. Dasgupta, D., Gonzalez, F., Yallapu, K., Gomez, J., Yarramsettii, R.: CIDS: An Agent-based Intrusion Detection System. Computers & Security 24(5), 387–398 (2005)
209. Cougaar: Cognitive Agent Architecture, `http://cougaar.org/`
210. Debar, H., Curry, D., Feinstein, B.: The Intrusion Detection Message Exchange Format (IDMEF). IETF RFC 4765 (2007)
211. Gowadia, V., Farkas, C., Valtorta, M.: PAID: A Probabilistic Agent-Based Intrusion Detection System. Computers & Security 24(7), 529–545 (2005)
212. Tsang, C.-H., Kwong, S.: Multi-agent Intrusion Detection System in Industrial Network using Ant Colony Clustering Approach and Unsupervised Feature Extraction. In: 2005 IEEE International Conference on Industrial Technology (ICIT 2005) (2005)
213. Mukkamala, S., Sung, A.H., Abraham, A.: Hybrid Multi-agent Framework for Detection of Stealthy Probes. Applied Soft Computing 7(3), 631–641 (2007)
214. Jansen, W.A., Karygiannis, T., Marks, D.G.: Applying Mobile Agents to Intrusion Detection and Response. US Department of Commerce, Technology Administration, National Institute of Standards and Technology (1999)
215. Asaka, M., Taguchi, A., Goto, S.: The Implementation of IDA: An Intrusion Detection Agent System. In: 11th Annual Computer Security Incident Handling Conference (June 1999)
216. De Queiroz, J.D., da Costa Carmo, L.F.R., Pirmez, L.: Micael: An Autonomous Mobile Agent System to Protect New Generation Networked Applications. In: Second International Workshop on Recent Advances in Intrusion Detection, RAID 1999 (1999)
217. Mell, P., Marks, D., McLarnon, M.: A Denial-of-service Resistant Intrusion Detection Architecture. Computer Networks: The International Journal of Computer and Telecommunications Networking 34(4), 641–658 (2000)
218. Krügel, C., Toth, T., Kirda, E.: SPARTA: a Mobile Agent Based Instrusion Detection System. In: IFIP TC11 WG11.4 First Annual Working Conference on Network Security: Advances in Network and Distributed Systems Security. IFIP Conference Proceedings, vol. 206. Kluwer, Dordrecht (2001)
219. Dasgupta, D., Brian, H.: Mobile Security Agents for Network Traffic Analysis. In: DARPA Information Survivability Conference & Exposition II, DISCEX 2001 (February 2001)
220. Helmer, G., Wong, J.S.K., Honavar, V.G., Miller, L.: Automated Discovery of Concise Predictive Rules for Intrusion Detection. Journal of Systems and Software 60(3), 165–175 (2002)
221. Helmer, G., Wong, J.S.K., Honavar, V., Miller, L., Wang, Y.: Lightweight Agents for Intrusion Detection. Journal of Systems and Software 67(2), 109–122 (2003)
222. Li, C., Song, Q., Zhang, C.: MA-IDS Architecture for Distributed Intrusion Detection using Mobile Agents. In: 2nd International Conference on Information Technology for Application, ICITA 2004 (2004)

223. Marks, D.G., Mell, P., Stinson, M.: Optimizing the Scalability of Network Intrusion Detection Systems Using Mobile Agents. Journal of Network and Systems Management 12(1), 95–110 (2004)
224. Deeter, K., Singh, K., Wilson, S., Filipozzi, L., Vuong, S.T.: APHIDS: A mobile agent-based programmable hybrid intrusion detection system. In: Karmouch, A., Korba, L., Madeira, E.R.M. (eds.) MATA 2004. LNCS, vol. 3284, pp. 244–253. Springer, Heidelberg (2004)
225. Alam, M.S., Gupta, A., Wires, J., Vuong, S.T.: APHIDS++: Evolution of A programmable hybrid intrusion detection system. In: Magedanz, T., Karmouch, A., Pierre, S., Venieris, I.S. (eds.) MATA 2005. LNCS, vol. 3744, pp. 22–31. Springer, Heidelberg (2005)
226. Kolaczek, G., Pieczynska-Kuchtiak, A., Juszczyszyn, K., Grzech, A., Katarzyniak, R.P., Nguyen, N.T.: A mobile agent approach to intrusion detection in network systems. In: Khosla, R., Howlett, R.J., Jain, L.C. (eds.) KES 2005. LNCS (LNAI), vol. 3682, pp. 514–519. Springer, Heidelberg (2005)
227. Foukia, N.: IDReAM: Intrusion Detection and Response Executed with Agent Mobility Architecture and Implementation. In: Fourth International Joint Conference on Autonomous Agents and Multiagent Systems (AAMAS 2005). ACM, The Netherlands (2005)
228. Alim, A.S.A., Ismail, A.S., Ahmed, S.H.: IDSUDA: An Intrusion Detection System Using Distributed Agents. Journal of Computer Networks and Internet Research 5(1), 1–11 (2005)
229. Wang, H.Q., Wang, Z.Q., Zhao, Q., Wang, G.F., Zheng, R.J., Liu, D.X.: Mobile agents for network intrusion resistance. In: Shen, H.T., Li, J., Li, M., Ni, J., Wang, W. (eds.) APWeb Workshops 2006. LNCS, vol. 3842, pp. 965–970. Springer, Heidelberg (2006)
230. Ma, K.-L.: Visualization for Security. ACM SIGGRAPH Computer Graphics 38(4), 4–6 (2004)
231. Erbacher, R.F., Walker, K.L., Frincke, D.A.: Intrusion and Misuse Detection in Large-scale Systems. IEEE Computer Graphics and Applications 22(1), 38–47 (2002)
232. Corchado, E., Wu, X., Oja, E., Herrero, Á., Baruque, B. (eds.): HAIS 2009. LNCS, vol. 5572. Springer, Heidelberg (2009)
233. Medsker, L.R.: Hybrid Intelligent Systems. Kluwer Academic Publishers, Dordrecht (1995)
234. Ron, S., Frederic, A. (eds.): Connectionist-Symbolic Integration: From Unified to Hybrid Approaches. Lawrence Erlbaum Associates, Inc., Mahwah (1997)
235. Tran, C., Abraham, A., Jain, L.: Decision Support Systems using Hybrid Neurocomputing. Neurocomputing 61, 85–97 (2004)
236. Bridges, S.M., Vaughn, R.B.: Fuzzy Data Mining and Genetic Algorithms Applied to Intrusion Detection. In: National Information Systems Security Conference (NISSC) (2000)
237. Marin, J.A., Ragsdale, D., Surdu, J.: A Hybrid Approach to Profile Creation and Intrusion Detection. In: DARPA Information Survivability Conference and Exposition II (DISCEX 2001) (2001)
238. Botha, M., Solms, R.v., Perry, K., Loubser, E., Yamoyany, G.: The Utilization of Artificial Intelligence in a Hybrid Intrusion Detection System. In: 2002 Annual Research Conference of the South African Institute of Computer Scientists and Information Technologists on Enablement through Technology, Port Elizabeth, South Africa. ACM International Conference Proceedings Series, vol. 30 (2002)

239. Sindhu, S.S.S., Ramasubramanian, P., Kannan, A.: Intelligent multi-agent based genetic fuzzy ensemble network intrusion detection. In: Pal, N.R., Kasabov, N., Mudi, R.K., Pal, S., Parui, S.K. (eds.) ICONIP 2004. LNCS, vol. 3316, pp. 983–988. Springer, Heidelberg (2004)
240. Middlemiss, M., Dick, G.: Feature Selection of Intrusion Detection Data Using a Hybrid Genetic Algorithm/KNN Approach. In: Design and Application of Hybrid Intelligent Systems, pp. 519–527. IOS Press, Amsterdam (2003)
241. Mukkamala, S., Sung, A.H., Abraham, A.: Intrusion Detection Using an Ensemble of Intelligent Paradigms. Journal of Network and Computer Applications 28(2), 167–182 (2005)
242. Kholfi, S., Habib, M., Aljahdali, S.: Best Hybrid Classifiers for Intrusion Detection. Journal of Computational Methods in Science and Engineering 6(2), 299–307 (2006)
243. Toosi, A.N., Kahani, M.: A New Approach to Intrusion Detection Based on an Evolutionary Soft Computing Model using Neuro-Fuzzy Classifiers. Computer Communications 30(10), 2201–2212 (2007)
244. Tsang, C.-H., Kwong, S., Wang, H.: Genetic-Fuzzy Rule Mining Approach and Evaluation of Feature Selection Techniques for Anomaly Intrusion Detection. Pattern Recognition 40(9), 2373–2391 (2007)
245. Peddabachigari, S., Abraham, A., Grosan, C., Thomas, J.: Modeling Intrusion Detection System Using Hybrid Intelligent Systems. Journal of Network and Computer Applications 30(1), 114–132 (2007)
246. Abadeh, M.S., Habibi, J., Lucas, C.: Intrusion Detection using a Fuzzy Genetics-based Learning Algorithm. Journal of Network and Computer Applications 30(1), 414–428 (2007)
247. Yuehui, C., Ajith, A., Bo, Y.: Hybrid Flexible Neural-tree-based Intrusion Detection Systems. International Journal of Intelligent Systems 22(4), 337–352 (2007)
248. Shon, T., Moon, J.: A Hybrid Machine Learning Approach to Network Anomaly Detection. Information Sciences 177(18), 3799–3821 (2007)
249. Goldring, T.: Scatter (and Other) Plots for Visualizing User Profiling Data and Network Traffic. In: 2004 ACM Workshop on Visualization and Data Mining for Computer Security. ACM Press, Washington (2004)
250. Labib, K., Vemuri, V.R.: Detecting and Visualizing Denial-of-Service and Network Probe Attacks Using Principal Component Analysis. In: Third Conference on Security and Network Architectures, SAR 2004 (2004)
251. Labib, K., Vemuri, V.R.: An Application of Principal Component Analysis to the Detection and Visualization of Computer Network Attacks. Annals of Telecommunications 61(1-2), 218–234 (2006)
252. Jianqiang, X., Dickerson, J.E., Dickerson, J.A.: Fuzzy Feature Extraction and Visualization for Intrusion Detection. In: 12th IEEE International Conference on Fuzzy Systems, vol. 2 (2003)
253. GGobi, http://www.ggobi.org/
254. Taylor, C., Alves-Foss, J.: NATE - Network Analysis of Anomalous Traffic Events, A Low-Cost Approach. In: New Security Paradigms Workshop (2001)
255. Shyu, M.L., Chen, S.C., Sarinnapakorn, K., Chang, L.: A Novel Anomaly Detection Scheme Based on Principal Component Classifier. In: IEEE Foundations and New Directions of Data Mining Workshop, pp. 172–179 (2003)
256. Lakhina, A., Papagiannaki, K., Crovella, M., Diot, C., Kolaczyk, E.D., Taft, N.: Structural Analysis of Network Traffic Flows. In: Joint International Conference on Measurement and Modeling of Computer Systems. ACM Press, New York (2004)

257. Bouzida, Y., Gombault, S.: Eigenconnections to Intrusion Detection. In: 19th IFIP International Information Security Conference. IFIP International Federation for Information Processing, vol. 147. Springer, Boston (2004)
258. Kuchimanchi, G.K., Phoha, V.V., Balagani, K.S., Gaddam, S.R.: Dimension Reduction using Feature Extraction Methods for Real-time Misuse Detection Systems. In: Fifth Annual IEEE SMC Information Assurance Workshop (2004)
259. Kurosawa, S., Nakayama, H., Nei, K., Jamalipour, A., Nemoto, Y.: A Self-adaptive Intrusion Detection Method for AODV-based Mobile Ad Hoc Networks. In: IEEE International Conference on Mobile Adhoc and Sensor Systems (2005)
260. Wang, W., Battiti, R.: Identifying Intrusions in Computer Networks with Principal Component Analysis. In: First International Conference on Availability, Reliability and Security, ARES 2006 (2006)
261. Ramah, K.H., Ayari, H., Kamoun, F.: Traffic anomaly detection and characterization in the tunisian national university network. In: Boavida, F., Plagemann, T., Stiller, B., Westphal, C., Monteiro, E. (eds.) NETWORKING 2006. LNCS, vol. 3976, pp. 136–147. Springer, Heidelberg (2006)
262. Venkatachalam, V., Selvan, S.: Performance Comparison of Intrusion Detection System Classifiers using Various Feature Reduction Techniques. International Journal of Simulation 9(1), 30–39 (2008)
263. Corchado, E., Herrero, Á.: Neural Visualization of Network Traffic Data for Intrusion Detection. Applied Soft Computing (2010) (accepted with changes)
264. Case, J., Fedor, M.S., Schoffstall, M.L., Davin, C.: Simple Network Management Protocol (SNMP). IETF RFC 1157 (1990)
265. Davin, J., Galvin, J., McCloghrie, K.: SNMP Administrative Model. IETF RFC 1351 (1992)
266. Vulnerability Statistics Report. Cisco Secure Consulting (2000)
267. Case, J., McCloghrie, K., Rose, M., Waldbusse, S.: Introduction to Version 2 of the Internet-standard Network Management Framework. IETF RFC 1441 (1993)
268. The Top 10 Most Critical Internet Security Threats (2000-2001 Archive). SANS Institute (2001)
269. Northcutt, S., Cooper, M., Fredericks, K., Fearnow, M., Riley, J.: Intrusion Signatures and Analysis. New Riders Publishing, Thousand Oaks (2001)
270. Myerson, J.M.: Identifying Enterprise Network Vulnerabilities. International Journal of Network Management 12(3), 135–144 (2002)
271. Mauro, D.R., Schmidt, K.J.: Essential SNMP. O'Reilly Media, Inc., Sebastopol (2001)
272. Perkins, D.T.: SNMP Versions. The Simple Times 5(1), 13–14 (1997)
273. Blumenthal, U., Wijnen, B.: Security Features of SNMPv3. The Simple Times 5(1), 8–12 (1997)
274. SNMP MIB, `http://edocs.bea.com/snmpagnt/v210/mibref/1tmib.html`
275. McCloghrie, K., Rose, M.: Management Information Base for Network Management of TCP/IP-based Internets: MIB-II. IETF RFC 1213 (1991)
276. Postel, J.: IAB Official Protocol Standards. IETF RFC 1100 (1989)
277. Staniford, S., Hoagland, J.A., McAlerney, J.M.: Practical Automated Detection of Stealthy Portscans. Journal of Computer Security 10(1-2), 105–136 (2002)
278. Malowidzki, M.: GetBulk Worth Fixing. The Simple Times 10(1), 3–6 (2002)

279. Sprenkels, R., Martin-Flatin, J.P.: Bulk Transfers of MIB Data. Technical Report SSC/1999/009. Communication Systems Division. Swiss Federal Institute of Technology Lausanne (1999)
280. Herrero, Á., Corchado, E., Pellicer, M.A., Abraham, A.: MOVIH-IDS: A Mobile-Visualization Hybrid Intrusion Detection System. Neurocomputing 72(13-15), 2775–2784 (2009)
281. Carrascosa, C., Bajo, J., Julián, V., Corchado, J.M., Botti, V.: Hybrid Multi-agent Architecture as a Real-Time Problem-Solving Model. Expert Systems with Applications: An International Journal 34(1), 2–17 (2008)
282. Bratman, M.E.: Intentions, Plans and Practical Reason. Harvard University Press, Cambridge (1987)
283. Herrero, Á., Corchado, E., Gastaldo, P., Zunino, R.: Neural Projection Techniques for the Visual Inspection of Network Traffic. Neurocomputing 72(16-18), 3649–3658 (2009)
284. Babu, S., Subramanian, L., Widom, J.: A Data Stream Management System for Network Traffic Management. In: Workshop on Network-Related Data Management (NRDM 2001) (2001)
285. Herrero, Á., Corchado, E.: Traffic data preparation for a hybrid network IDS. In: Corchado, E., Abraham, A., Pedrycz, W. (eds.) HAIS 2008. LNCS (LNAI), vol. 5271, pp. 247–256. Springer, Heidelberg (2008)
286. Dreger, H., Feldmann, A., Paxson, V., Sommer, R.: Operational Experiences with High-Volume Network Intrusion Detection. In: 11th ACM Conference on Computer and Communications Security. ACM Press, New York (2004)
287. Wireshark, `http://www.wireshark.org`
288. Cisco IOS NetFlow, `http://www.cisco.com/web/go/netflow`
289. Corchado, J.M., Laza, R.: Constructing Deliberative Agents with Case-Based Reasoning Technology. International Journal of Intelligent Systems 18(12), 1227–1241 (2003)
290. Pellicer, M.A., Corchado, J.M.: Development of CBR-BDI Agents. International Journal of Computer Science and Applications 2(1), 25–32 (2005)
291. Bajo, J., Corchado, J.M., Rodríguez, S.: Intelligent guidance and suggestions using case-based planning. In: Weber, R.O., Richter, M.M. (eds.) ICCBR 2007. LNCS (LNAI), vol. 4626, pp. 389–403. Springer, Heidelberg (2007)
292. Hammond, K.J.: Case-based Planning: Viewing Planning as a Memory Task. Academic Press Professional, Inc., London (1989)
293. Spalzzi, L.: A Survey on Case-Based Planning. Artificial Intelligence Review 16(1), 3–36 (2001)
294. Zambonelli, F., Jennings, N.R., Wooldridge, M.: Developing Multiagent Systems: the Gaia Methodology. ACM Transactions on Software Engineering and Methodology 12(3), 317–370 (2003)
295. Bevilacqua, A.: A Dynamic Load Balancing Method on A Heterogeneous Cluster of Workstations. Informatica 23(1) (1999)
296. Schaerf, A., Shoham, Y., Tennenholtz, M.: Adaptive Load Balancing: A Study in Multi-Agent Learning. Journal of Artificial Intelligence Research 2, 475–500 (1995)
297. Noronha Nassif, L., Marcos Nogueira, J., Vinicius de Andrade, F., Ahmed, M., Karmouch, A., Impey, R.: Job Completion Prediction in Grid using Distributed Case-based Reasoning. In: 14th IEEE International Workshop on Enabling Technologies: Infrastructure for Collaborative Enterprise (2005)

298. Mahoney, M.V., Chan, P.K.: An analysis of the 1999 dARPA/Lincoln laboratory evaluation data for network anomaly detection. In: Vigna, G., Krügel, C., Jonsson, E. (eds.) RAID 2003. LNCS, vol. 2820, pp. 220–237. Springer, Heidelberg (2003)
299. McHugh, J.: Testing Intrusion Detection Systems: A Critique of the 1998 and 1999 Darpa Off-line Intrusion Detection System Evaluation as Performed by Lincoln Laboratory. ACM Transactions on Information and System Security 3(4), 262–294 (2000)
300. DARPA Intrusion Detection Evaluation, `http://www.ll.mit.edu/mission/communications/ist/corpora/ideval/index.html`
301. García-Teodoro, P., Díaz-Verdejo, J., Maciá-Fernández, G., Vázquez, E.: Anomaly-based Network Intrusion Detection: Techniques, Systems and Challenges. Computers & Security 28(1-2), 18–28 (2009)
302. Ranum, M.J.: Experiences Benchmarking Intrusion Detection Systems. NFR Security Technical Publications (2001)
303. Egan, J.P.: Signal Detection Theory and ROC-analysis. Series in Cognition and Perception. Academic Press, London (1975)
304. Mueller, P., Shipley, G.: Dragon Claws its Way to the Top. Network Computing 20, 45–67 (2001)
305. Athanasiades, N., Abler, R., Levine, J., Owen, H., Riley, G.: Intrusion Detection Testing and Benchmarking Methodologies. In: First IEEE International Workshop on Information Assurance (2003)
306. Maheshkumar, S., Gursel, S.: Why Machine Learning Algorithms Fail in Misuse Detection on KDD Intrusion Detection Data Set. Intelligent Data Analysis 8(4), 403–415 (2004)
307. Brugger, S.T., Chow, J.: An Assessment of the DARPA IDS Evaluation Dataset Using Snort. Technical Report. Department of Computer Science - UC Davis (2007)
308. Bermúdez-Edo, M., Salazar-Hernández, R., Díaz-Verdejo, J.E., García-Teodoro, P.: Proposals on assessment environments for anomaly-based network intrusion detection systems. In: López, J. (ed.) CRITIS 2006. LNCS, vol. 4347, pp. 210–221. Springer, Heidelberg (2006)
309. Haines, J.W., Rossey, L.M., Lippmann, R.P., Cunningham, R.: Extending the DARPA Off-Line Intrusion Detection Evaluations. In: DARPA Information Survivability Conference and Exposition II, vol. 1 (2001)
310. Alessandri, D.: Using rule-based activity descriptions to evaluate intrusion-detection systems. In: Debar, H., Mé, L., Wu, S.F. (eds.) RAID 2000. LNCS, vol. 1907, p. 183. Springer, Heidelberg (2000)
311. Mueller, P., Shipley, G.: To Catch a Thief. Network Computing 20 (2001)
312. Newman, D., Snyder, J., Thayer, R.: Crying Wolf: False Alarms Hide Attacks. Network World (2002)
313. Network Intrusion Prevention System Tests, `http://nsslabs.com/ips`
314. Maxion, R.A., Tan, K.M.C.: Benchmarking Anomaly-based Detection Systems. In: International Conference on Dependable Systems and Networks (DSN 2000) (2000)
315. Mell, P., Hu, V., Lippmann, R.: An Overview of Issues in Testing Intrusion Detection Systems. NIST Interagency Reports. National Institute of Standards and Technology - Information Technology Laboratory (2003)
316. Vigna, G., Robertson, W., Balzarotti, D.: Testing Network-Based Intrusion Detection Signatures Using Mutant Exploits. In: 11th ACM Conference on Computer and Communications Security. ACM Press, Washington (2004)
317. Marti, R.: THOR: A Tool to Test Intrusion Detection Systems by Variations of Attacks. Diploma Thesis. ETH Zurich (2002)

318. Herrero, Á., Corchado, E., Gastaldo, P., Zunino, R.: A comparison of neural projection techniques applied to intrusion detection systems. In: Sandoval, F., Prieto, A.G., Cabestany, J., Graña, M. (eds.) IWANN 2007. LNCS, vol. 4507, pp. 1138–1146. Springer, Heidelberg (2007)
319. SOM Toolbox for Matlab, `http://www.cis.hut.fi/projects/somtoolbox/`
320. Goodall, J.R., Lutters, W.G., Rheingans, P., Komlodi, A.: Preserving the Big Picture: Visual Network Traffic Analysis with TNV. In: IEEE Workshop on Visualization for Computer Security (VizSEC 2005). IEEE Computer Society, Los Alamitos (2005)
321. Mukkamala, S., Janoski, G., Sung, A.: Intrusion Detection Using Neural Networks and Support Vector Machines. In: 2002 International Joint Conference on Neural Networks, IJCNN 2002 (February 2002)
322. Qiao, Y., Xin, X.W., Bin, Y., Ge, S.: Anomaly Intrusion Detection Method Based on HMM. Electronics Letters 38(13), 663–664 (2002)
323. Liao, Y.H., Vemuri, V.R.: Use of K-Nearest Neighbor Classifier for Intrusion Detection. Computers & Security 21(5), 439–448 (2002)
324. Yi, M.K., Hwang, C.S.: Intrusion-tolerant intrusion detection system. In: Chen, H., Moore, R., Zeng, D.D., Leavitt, J. (eds.) ISI 2004. LNCS, vol. 3073, pp. 476–483. Springer, Heidelberg (2004)
325. Shneiderman, B.: The Eyes Have It: a Task by Data Type Taxonomy for Information Visualizations. In: IEEE Symposium on Visual Languages (1996)

GPSR Compliance

The European Union's (EU) General Product Safety Regulation (GPSR) is a set of rules that requires consumer products to be safe and our obligations to ensure this.

If you have any concerns about our products, you can contact us on ProductSafety@springernature.com

In case Publisher is established outside the EU, the EU authorized representative is:

Springer Nature Customer Service Center GmbH
Europaplatz 3
69115 Heidelberg, Germany

Batch number: 10614940

Printed by Printforce, the Netherlands